Animal Cognition

WITHDRAWN

Animal Cognition looks at how non-human animals process information from their environment. Nick Lund has written an accessible and engaging account of this area of comparative psychology. The book contains chapters that range from navigation and memory to communication methods, and attempts to teach language to non-human animals.

Nick Lund is a Senior Lecturer in Psychology at Manchester Metropolitan University.

Routledge Modular Psychology

Series editors: Cara Flanagan is a Reviser for AS and A2 level Psychology and lectures at Inverness College. Philip Banyard is Associate Senior Lecturer in Psychology at Nottingham Trent University and a Chief Examiner for AS and A2 level Psychology. Both are experienced writers.

The *Routledge Modular Psychology* series is a completely new approach to introductory-level psychology, tailor-made to the new modular style of teaching. Each short book covers a topic in more detail than any large textbook can, allowing teacher and student to select material exactly to suit any particular course or project.

The books have been written especially for those students new to higher-level study, whether at school, college or university. They include specially designed features to help with technique, such as a model essay at an average level with an examiner's comments to show how extra marks can be gained. The authors are all examiners and teachers at the introductory level.

The *Routledge Modular Psychology* texts are all user-friendly and accessible and include the following features:

- practice essays with specialist commentary to show how to achieve a higher grade
- chapter summaries to assist with revision
- progress and review exercises
- glossary of key terms
- summaries of key research
- further reading to stimulate ongoing study and research
- cross-referencing to other books in the series

Also available in this series (titles listed by syllabus section):

ATYPICAL DEVELOPMENT AND ABNORMAL BEHAVIOUR

Psychopathology
John D. Stirling and Jonathan S.E. Hellewell

Therapeutic Approaches in Psychology
Susan Cave

Classification and Diagnosis of Psychological Abnormality
Susan Cave

BIO-PSYCHOLOGY

Cortical Functions
John Stirling

The Physiological Basis of Behaviour: Neural and hormonal processes
Kevin Silber

Awareness: Biorhythms, sleep and dreaming
Evie Bentley

COGNITIVE PSYCHOLOGY

Memory and Forgetting
John Henderson

Perception: Theory, development and organisation
Paul Rookes and Jane Willson

Attention and Pattern Recognition
Nick Lund

DEVELOPMENTAL PSYCHOLOGY

Early Socialisation: Sociability and attachment
Cara Flanagan

Social and Personality Development
Tina Abbott

PERSPECTIVES AND RESEARCH

Controversies in Psychology
Philip Banyard

Ethical Issues and Guidelines in Psychology
Cara Flanagan and Philip Banyard (forthcoming)

Introducing Research and Data in Psychology: A guide to methods and analysis
Ann Searle

Theoretical Approaches in Psychology
Matt Jarvis

Debates in Psychology
Andy Bell

SOCIAL PSYCHOLOGY

Social Influences
Kevin Wren

Interpersonal Relationships
Diana Dwyer

Social Cognition
Donald C. Pennington

COMPARATIVE PSYCHOLOGY

The Determinants of Animal Behaviour
Jo-Anne Cartwright

Evolutionary Explanations of Human Behaviour
John Cartwright

OTHER TITLES

Sport Psychology
Matt Jarvis

Health Psychology
Anthony Curtis

Psychology and Work
Christine Hodson

Psychology and Education
Susan Bentham (forthcoming)

Psychology and Crime
David Putwain and Aidan Sammons

STUDY GUIDE

Exam Success in AQA(A) Psychology
Paul Humphreys (forthcoming)

To Sue

Animal Cognition

Nick Lund

First published 2002 by Routledge
27 Church Road, Hove, East Sussex, BN3 2FA

Simultaneously published in the USA and Canada
by Taylor & Francis Inc.
29 West 35th Street, New York, NY 10001

Routledge is an imprint of the Taylor & Francis Group

© 2002 Psychology Press

Typeset in Times and Frutiger by Keystroke,
Jacaranda Lodge, Wolverhampton
Printed and bound in Great Britain
by TJ International, Padstow, Cornwall

Cover design by Terry Foley

British Library Cataloguing in Publication Data
A catalogue record for this book is available from the British Library

Library of Congress Cataloging-in-Publication Data
Lund, Nick, 1956–
 Animal cognition / Nick Lund.
  p. cm. – (Routledge modular psychology)
 Includes bibliographical references (p.).
 ISBN 0-415-25298-9 (pbk.) – ISBN 0-415-25297-0 (hbk.)
 1. Cognition in animals. I. Title. II. Series.

QL785 .L795 2002
591.5—dc21   2001056892

ISBN 0–415–25297–0 (hbk)
ISBN 0–415–25298–9 (pbk)

Contents

Figures

Introduction to animal cognition

Introduction

When I was young my parents took me to a zoo. When we arrived at the chimpanzee enclosure I noticed a young male chimpanzee looking at me so I put my hand in the air to wave at him. He also put his hand in the air. I then walked beside the barrier separating me from the enclosure. He walked along the inside of the enclosure. I turned and ran back, and so did he. I then went from one end to the other by jumping. The chimpanzee jumped with me. For the next 20 minutes or so the chimpanzee mimicked all of my actions and even some of my facial expressions. Sometimes the chimpanzee led the activity and I copied him. What was the chimpanzee *thinking*? Did the chimpanzee's actions indicate mental process such as *short-term memory*? What information did the chimpanzee need to process in order to mimic my actions?

I have a friend who has a cat that opens doors by jumping up and pulling on the door handle. Does the cat *remember* how to open doors? Does it have a *mental map* of the house?

These are all questions about animal cognition, or how animals process information. Most of us have some anecdotal evidence that seems to indicate some form of mental ability in animals (there have even been reports of pigeons riding on underground trains to find food in London!). The purpose of this book is to look at the *science* of animal cognition and to examine the evidence of cognitive abilities.

What is 'animal cognition'?

The study of human cognition is a very important part of psychology. Psychologists who use the 'cognitive approach' view people as complicated processors of information. We are constantly bombarded by information. We tend to ignore some of the stimuli and pay attention to others. Some of the stimuli we pay attention to we quickly forget and some is stored for long periods to be used later. Cognitive psychologists study human behaviour and responses to try to analyse the steps in the processing of the information. In the past 20–30 years this approach has been applied to non-human animals. They too are constantly bombarded with information that they need to analyse and process in order to survive. It is probable that animals, like humans, selectively attend to some stimuli, perceive patterns, store information for future use and communicate information to other animals. Animals do not merely respond to information; like humans they process it.

The term 'cognition' has caused some confusion when applied to animal behaviour because different people use it in slightly different ways. Dawkins (1995) describes how the term 'cognitive' is usually used to refer to any animal behaviour that requires the processing of information, but for some the term 'cognitive' has become tied up with the issue of consciousness to such an extent that the terms are synonymous.

To illustrate the idea of information processing Pearce (1997) uses the example of an animal that is repeatedly presented with a tone and food. It is assumed that this results in the 'formation of internal, central representations of the tone and food' (p. 15). The task for anyone interested in animal cognition is to study how these representations are encoded and how the relationship between them is established. Pearce points out:

> In a sense a central representation of an environmental event constitutes an essential component of animal cognition. The task confronting a person interested in this topic is to show what these representations consist of and how they function in the higher mental processes.
>
> (Pearce, 1997, p. 15)

In other words, the cognitive approach assumes there is something going on in the animal's brain which corresponds to the outside world and that the animal is able to respond to this information appropriately.

To other researchers the term 'cognitive' refers to mental processes that imply consciousness, since humans are aware of these mental processes. For example, if you get up to go to the kitchen to get a drink you are aware of where you are going and why. Later on you will also be aware of the memory of getting the drink. Imagine a bird flying from its nest to a bird table where it regularly finds food: it navigates to the table, it appears to have a representation of the table and it seems to have developed an association between the table and food. The question is, does this imply an awareness or consciousness of self? For some time researchers of animal behaviour resisted the notion of animal consciousness. However, in his influential book *Animal Thinking*, Griffin (1984) argued that it was not possible to deny the existence of thought processes and even consciousness in some animals. He acknowledges that it is difficult to study thought and consciousness in non-human animals, but this does not mean they do not exist. He points out that it is hard to believe that mind and consciousness should have suddenly evolved for humans only. From an evolutionary point of view we should expect some precursors to human thought in other animals. Others have argued that the complex interactions found in animals like the social primates require thought and may have acted as a powerful evolutionary force (Humphrey, 1976). For example, if an animal could not recognise and remember which members of the group were dominant and which were subordinate then they would not survive in the group for long.

However, it is important to note that cognitive processes (or **cognition**) are not the same as consciousness. Like Pearce, Roberts (1998) suggests that the study of animal cognition looks at models of

3

how animals form representations about their environment and then process information. These models are then used to try to organise, understand and predict animal behaviour. However, he points out: 'This discussion proceeds without a consideration of consciousness because we have no idea of the processes an animal might or might not be aware of and whether such awareness would be of any functional consequence' (p. 17).

This book is concerned with animal *cognition* or the *processing of information*; the topics of self-awareness and theory of mind (which relate to consciousness) are dealt with in another title in the series *Determinants of Animal Behaviour* (Cartwright, 2002).

For convenience the term 'animal' will be used to refer to non-human animals in the rest of the book, except when it is important that the difference between non-human animals and humans needs to be emphasised.

Navigation, communication and memory

This book has been organised into three broad areas to investigate some mental processes in animals: navigation, communication and memory. For convenience these topics are presented as if they are discrete aspects of animal cognition. In reality of course this is not the case. A wasp that uses landmarks to find a burrow is demonstrating memory and is navigating. A chimpanzee that uses a symbol to indicate an orange is demonstrating both communication and memory. Many of the studies cited in this book can therefore be used in a variety of essays. For example, the famous studies of the honey-bee dance by von Frisch (see key summary 1 on pp. 113–116) can be used in essays on communication (the dance indicates that there is a food source), memory (the dance is about something the bee encountered some time ago) and navigation (the dance indicates where a food source is to be located).

Navigation

A common feature of animals is that they move around in their environment. A lot of this movement seems to be directed; animals move to food sources, home or breeding grounds. Movement that is directed to a particular goal requires **navigation** and navigation needs some form of representation of the external environment.

Some animals use navigation regularly in order to return to shelter after feeding, foraging, etc. This is called **homing** behaviour. A different scale of navigation is needed for long-distance travel that some animals make on a seasonal basis, or **migration**. Navigation in homing behaviour is discussed in Chapter 2 and navigation in migration in Chapter 3.

Communication and language

Many animals have some form of communication system and at its simplest this is the transfer of information from one animal to another. **Communication** is an important aspect of animal behaviour. For example, it would be impossible to live in a social group without some form of communication. Animals communicate using many different methods, ranging from the visual to the chemical. The differences reflect the diversity of animal species and habitats.

Communication in humans usually involves **language**. However, communication does not require language and the two are not synonymous. Language is a special and complex form of communication. Animal communication in non-human animals is discussed in Chapter 4. Whether language exists in non-human animals or whether it can be taught to non-human animals is discussed in Chapter 5.

Memory

A lot of animal behaviour seems to indicate the use of memory. For example, many animals seem to develop an intimate knowledge of their local environment, such as where to find different food sources, where their home is and where danger lies. This requires **spatial memory** (or **cognitive maps**). Some animals store food in order to survive times when little food is available, a behaviour known as **food caching**. In order for this to be useful the animal needs to remember where the food is stored. Memory in non-human animals is discussed in Chapter 6.

Problems of studying animal cognition

The study of human cognition is difficult because we cannot observe the mental processes that may occur. However, we can ask people what they perceive, remember or understand. It is more difficult to study animal cognition since we cannot talk to them to establish what they 'think' and, as with humans, there is no direct way of observing any mental processes (see pp. 10–11).

Apart from the difficulties of studying mental processes and communication three other problems can be identified: animal senses, animal diversity and **anthropomorphism**.

Animal senses

One of the problems of studying navigation, communication and memory in animals is that the sensory abilities of animals are not necessarily the same as humans. Some animals sense information that is beyond the capabilities of human senses. For example, pigeons can detect sounds as low as 0.06 Hz (called infrasound) which they may use in navigation (Yodlowski *et al.*, 1977) but humans cannot detect frequencies less than 10 Hz. Animals can detect sounds, odours, tastes and light that are undetected by humans. Some animals can detect stimuli which humans are not able to sense at all. For example, some fish detect objects around them by using an electric field (Lissman, 1963). A further problem is that there are stimuli that humans can detect that some animals cannot. Animals may also produce stimuli that go unnoticed by humans. For example, dolphins can produce sounds that are far higher in frequency than anything humans can hear.

The reason that this can pose a problem is that it is easy to assume that animals are responding to one stimulus when they are responding to another stimulus that is undetected by the human investigator. It is also easy to assume that an animal is responding to a stimulus because the human investigator can detect it, although the animal may not be able to. The study of animal behaviour therefore often requires specialist equipment in order to try to understand the world of the animal being studied and to give a full picture of the animal's behaviour.

Animal diversity

A further problem when studying animal cognition is the sheer diversity of behaviour and abilities that exist amongst animals. A complete understanding of the cognitive processes behind a behaviour of one species does not necessarily help explain a similar behaviour in a different species. For example, a thorough understanding of how a salmon returns to a stream to spawn does not help us to understand how a bee returns to its hive. The reason for the diversity of behaviour and cognitive processes is that there is great diversity in the demands made upon animals. One would not expect the same cognitive abilities in a solitary insect like the digger wasp as in a social mammal like the chimpanzee. The nature of the animal's environment also has a great impact on communication, navigation and memory. Animals live in a wide variety of environments: water, air, arboreal, land, subterranean, etc. An animal walking over land can use landmarks to navigate but an animal flying over an ocean cannot. This book cannot cover the cognitive abilities of all species (readers who have interest in a particular species or topic should refer to some of the specialist texts mentioned at the end of each chapter). In each chapter a range of examples have been chosen to illustrate the types of abilities and strategies that animals demonstrate.

Anthropomorphism

When studying the behaviour of animals it is tempting to describe the behaviour in human terms. If your cat rushes to you in the kitchen and starts rubbing against your legs and purring it is tempting to conclude that the cat is 'thinking' about food and is 'feeling' hungry. When one studies the complexities involved in communication or in navigating over vast distances it becomes difficult not to talk about 'thinking' or 'consciousness'. It can also be tempting to ascribe motives and feelings to the animal. However, these are human activities and attributes and it is difficult to show that animals are the same. Using human characteristics to describe and explain animal behaviour is called anthropomorphism. Study the example in Box 1.1.

Box 1.1 Clever Hans: the horse who could count (Pfungst, 1965)

One of the earliest studies of animal cognition was carried out in the early 1900s on 'Clever Hans', a horse. Hans's owner claimed that the horse could count and was capable of quite complex mathematics. When Hans was given a problem to solve he would respond by striking the ground with his hoof to indicate the answer. For example, when given a simple problem such as 'What is 4 plus 2?' Hans would strike his hoof on the ground six times. Hans also seemed to be able to perform subtractions, multiplications and divisions. There was even evidence that he could add fractions, something that confuses a lot of humans!

Understandably there was disbelief about these claims and a committee of 13 men investigated Hans. They found that the owner's claim was true and concluded that Hans could indeed count. Whenever his trainer set a problem Hans nearly always gave the right answer.

However, on closer examination by psychologists it became evident that Hans could not count. Instead he was responding to barely perceptible cues being given (unwittingly) by the trainer. When Hans reached the correct number of hoof strikes the trainer's expression changed slightly. Hans responded to this cue and stopped. When the 'right answer' had been given the trainer often gave Hans some food and therefore reinforced the horse's behaviour. When the trainer's face was hidden Hans did not give the correct answer.

Hans had a remarkable ability to respond to very subtle cues but he could not count.

The study of 'Clever Hans', and other examples, shows that anthropomorphism can lead to complete misinterpretation of an animal's behaviour. Such examples led one of the pioneers of comparative psychology, Lloyd Morgan (1894), to suggest that no animal behaviour should be explained in terms of higher mental processes (e.g. thinking, planning, etc.) if it could be explained by simpler processes (e.g. conditioned responses etc.). This became known as **Lloyd Morgan's canon**. In the case of Clever Hans the horse's behaviour was initially explained by higher mental processes (counting), but it was better explained by a simpler process (conditioned to respond to visual cues).

Psychologists and zoologists who study animal behaviour generally try hard to avoid anthropomorphism and to be objective. However, this can lead to a different problem. In striving to avoid anthropomorphism, are researchers ignoring evidence that animals do process information in the same way as humans and even have some form of consciousness? This is a topic that triggers many debates in comparative psychology and is unlikely to be resolved easily. Two different perspectives on the debate can be seen in the views of Kennedy (1992) and Fisher (1996). In his book *The New Anthropomorphism* Kennedy argues that anthropomorphism still exists in research into animal behaviour but that it now has a more subtle influence. Although scientists are careful to avoid blatant anthropomorphism, such as 'the chimpanzee was jealous and resentful', they still use terms which can suggest human attributes. For example, one term that is commonly used is 'optimal' (as in, optimal foraging theory). Describing an animal's behaviour as optimising chances of success (in finding food, a mate, etc.) is fine. However, Kennedy suggests that then it is easy to believe that the animal 'knows' that one act will probably be more successful than another, or that the animal can calculate odds. Thus some terms lead to an underlying assumption of human qualities in the animal.

On the other hand, Fisher (1996), in a chapter called 'The Myth of Anthropomorphism', argues that the dangers of anthropomorphism are overstated and that the constant pressure to explain behaviour in the simplest form leads us to disregard and underestimate the capabilities of animals. Fisher claims: 'The charge of anthropomorphism oversimplifies a complex issue – animal consciousness – and it tries to inhibit consideration of positions that ought to be evaluated in a more open-minded and empirical manner' (p. 3).

Progress exercise

In a series of famous studies Köhler (1925) investigated problem-solving in chimpanzees. The problem usually involved some fruit that was out of reach of the chimpanzee but which could be reached with the use of equipment that Köhler placed nearby. For example, in one test Köhler placed some fruit outside the cage of a chimpanzee. The fruit was deliberately placed beyond the chimpanzee's reach. Köhler placed a short stick outside the cage, but this was too short to reach the fruit. However, the short stick could be used to reach a longer stick which could reach the fruit. Köhler found that the chimpanzee did not get the fruit by trial and error but seemed to study the problem and then reached for the short stick, pulled the long stick to the cage and then used the long stick to get the fruit.

Write down what we can infer about mental processes from this example.

Examine what you have written carefully. Could the behaviour be explained in a simpler way?

The study of cognition in animals

As noted earlier, the main problem in studying animal cognition is that there is no way of observing mental processes directly. The only thing that can be studied directly is animal behaviour. The existence of any mental processes or representations has to be *inferred* from the behaviour. Thus the task for psychologists interested in animal cognition is to devise experiments and studies that can demonstrate mental processes through observation of animal behaviour.

Using animal behaviour to infer mental processes can be problematic. It is not always possible to devise a study that can demonstrate any mental process unambiguously. Often a behaviour can be explained in very different ways. For example, some chimpanzees have been taught to use sign language (see Chapter 5). One inference from this is that the chimpanzees have learned to use and understand 'language'. An alternative explanation of the chimpanzees' behaviour is that they have learned a series of conditioned responses to obtain food (i.e. they have no more concept of language than a rat pushing a lever to get food in a Skinner box).

How can we choose between conflicting explanations of behaviour? Pearce (1997) has suggested there are two ways of addressing the problem, the first based on the nature of theory and the second based on the experimental method. The theoretical approach to answering the problem relates to Lloyd Morgan's canon. This states that if a behaviour can be explained by a simple process or a more complex process we should always accept the simplest. In the example of the chimpanzees, if the evidence was ambiguous then we should accept the simplest explanation: that they have learned conditioned responses. The danger of this approach has already been mentioned: the assumption that the simplest explanation is correct may be wrong and we may ignore evidence of higher mental abilities.

The second approach is to use the experimental method. If the data from a study is ambiguous then the best way to choose between alternative explanations is to generate new hypotheses that can be tested by further experiments. For example, if we could devise a procedure in which the chimpanzees could demonstrate a novel use of signs that showed a use of grammatical rules then this would be an indication of some language ability. The problem with this solution is that it is difficult to devise studies that demonstrate a mental function unambiguously.

As Shettleworth (1998) notes 'formulating unambiguous alternatives does not guarantee finding unambiguous answers' (p. 573). Animals may show a cognitive ability in one situation but not in another. Alternatively, there may be evidence that animals show some attributes of a cognitive ability but not others (e.g. language is a complex ability involving many different attributes, animals may show some but not others). Allen and Hauser (1996) liken the problems of studying animal cognition to the problems of studying the cognitive development of pre-linguistic children. It is theoretically possible to devise experiments to demonstrate cognitive processes, but there are often practical or ethical difficulties involved.

Recently an interesting alternative to studying animal responses has emerged: animal robotics (see for example, McFarland, 1999). This approach attempts to build models (or robots) which act autonomously and are designed to act like animals. These animal-like robots (often referred to as **animats**) are machines which process information in a precise, predetermined way. If an animat can be designed to mimic an animal's behaviour precisely we can identify what information is needed and how it has to be processed. For example, if we can design an animat to 'learn' where there is a good 'food source' it seems reasonable to assume that an animal could learn in the same way.

Shettleworth (1998) points out that there have been two approaches to the study of animal cognition. One approach, which she describes as the *ecological program*, studies animal cognition in terms of ecology and evolution. In other words, what selective pressures led to the animal's abilities, how are they useful in the animal's environment, etc. The other approach, the *traditional program*, is more **anthropocentric** in emphasis and tends to be interested in how animals perform similar cognitive tasks to humans (e.g. do animals have cognitive maps? How do animals communicate?). Shettleworth (1998) believes there are signs of a synthesis between these two approaches and that new questions are being asked about animal cognition. The study of animal cognition seems to be at an exciting stage of development.

Review exercise

Review the following account of animal behaviour and note:

(a) any cognitive abilities that are demonstrated;
(b) any evidence of anthropomorphism in my account.

Orang-utans have the reputation as escape artists in zoos. In his very entertaining and interesting book, *The Parrot's Lament*, Eugene Linden describes a number of escapes by orang-utans. Some of these involved the use of tools that the orang-utans kept hidden from their keepers. One ingenious male escaped several times using a length of wire that he kept in his mouth. Most of the time the apes did not go far but sat near to their enclosure apparently enjoying the consternation they caused. One orang-utan was temporarily kept isolated because of his aggressive behaviour and he seemed to dislike it. Waiting patiently until an opportunity occurred, he eventually used both his great strength and a tool made from cardboard to overcome a series of security measures. Once out he got a bucket and mop and began cleaning the floor!

Summary

The study of animal cognition is the study of how animals process and respond to information from the environment. Animals need to process information in many aspects of their lives, but three behaviours in which information processing is essential are navigation, communication and memory. Navigation involves using some information about the environment to guide movement to a particular goal. Communication is important to all animals and involves the transfer of information from one animal to another. Many animal activities require the animal to store information and this storing of information, or memory, is vital to the survival of most animals. The study of animal cognition is difficult because it is the study of internal processes. This is further complicated by the differences in animal senses and the diversity of animal behaviour and abilities. There is also the danger of animal behaviour being explained in human terms (anthropomorphism). Cognitive abilities in animals cannot be observed directly and can only be inferred from behaviour. Studies of animal cognition need to be designed rigorously so that the behaviour demonstrates the cognitive ability unambiguously.

Further reading

Pearce, J.M. (1997) *Animal Learning and Cognition* (2nd edn), Hove: Psychology Press. This is an advanced textbook but it is easy to read since it presents ideas and information very clearly. It deals with the topics covered in the 'Animal Cognition' section of the AQA syllabus and, in addition, it has excellent sections on learning and social learning in non-human animals (useful also for the 'Determinants of Animal Behaviour' section).

Shettleworth, S.J. (1998) *Cognition, Evolution and Behaviour*, New York: Oxford University Press. This is an excellent textbook for anyone with an interest in animal cognition. It is not aimed at A-level students but it is clear and detailed.

Students who enjoy reading about the remarkable abilities of animals could look at Professor Linden's book that was mentioned in the review exercise: Linden, E. (2000) *The Parrots Lament: And Other True Tales of Animal Intrigue, Intelligence and Ingenuity*, London: Souvenir Press Ltd.

2

Animal navigation: navigational techniques and homing behaviour

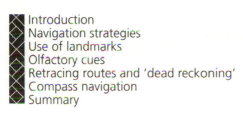
Introduction

A typical feature of animals is that they move. Some animal movement may be haphazard but frequently animals move to precise locations to find food, to a water supply, to a suitable environment to raise young, to return home, etc. Some animals only move a few metres during a lifetime whilst others can cover vast distances. Whatever the distance, movement to a destination or goal requires navigation.

Waterman (1989) points out that a mobile animal's habitat usually consists of a number of subhabitats and that to move between them effectively the animal must travel in the right direction, to the right distance and at the right time. The purpose of the next two chapters is to investigate the cognitive mechanisms that enable them to achieve this. Many animals have a base of some sort such as a nest, a burrow or a cave. When animals leave this base to find food and water they need a way of finding it again; this use of navigation to return to a base is termed homing behaviour. This chapter will concentrate on the navigation aids used for the relatively short-distance travel used in homing behaviour. Homing behaviour is directed at a specific goal and therefore navigation has to be precise. Chapter 3 concentrates on navigation over long distances and is concerned with migration. Unlike homing behaviour it is not necessarily directed to a precise location but often to a general area.

Some journeys may involve both migration and homing. For example, a swallow returning from Africa to its old nest in England is both migrating and, at the end of the journey, showing homing behaviour. In such cases the strategies used to navigate during the long-distance travel (migration) are likely to be different to finding a nest or shelter (homing).

Navigational strategies

There are a number of cues that can be used to navigate through an environment. Animals can use landmarks, the sun, stars, odours and even magnetic fields as guides. Animals vary widely in their ability to navigate, and different species use different cues or combinations of cues in their homing behaviour. Furthermore, one species may well use different strategies depending on the circumstances and/or the conditions they are experiencing. For example, pigeons use a number of different cues to navigate, including magnetic compass, sun compass, odours, ambient pressure and landmarks. The importance of each cue varies, depending on the conditions (a sun compass cannot be used on a very overcast day) and distance from the loft (landmarks cannot be seen from great distances).

McFarland (1999) has identified three types of orientation that are important in navigation:

1 *Pilotage*: navigation using familiar landmarks or features of some sort (visual, olfactory, etc.). For example, some insects navigate towards objects that are associated with their nest. If the objects are moved the insect navigates towards the new location.
2 *Compass orientation*: navigation using a particular compass direction without using landmarks. For example, Baker (1978) has reported that the small white butterfly moves in the same direction regardless of wind direction day after day.

3 *True navigation*: the ability to navigate to a goal point without the use of landmarks and regardless of the direction. Pigeons seem to use true navigation and can return to their loft no matter where they are released.

One method for demonstrating the difference between **compass orientation** and true navigation is to use displacement of animals during a journey (Figure 2.1). Animals are captured in one location and released in another. If an animal then proceeds in the same direction without compensating when it is released then it is using compass orientation (Figure 2.1a). However, if an animal corrects for the displacement on release and changes direction to head for the original destination it is evidence that it is using **true navigation** (Figure 2.1b).

In the following discussion of different navigational techniques it is evident that all these types of orientation are used. The use of landmarks, for example, involves **pilotage**; dead reckoning, on the other hand, seems to use compass orientation. The 'compass' techniques (using the earth's magnetic field, the sun or stars) can be used for either compass orientation (e.g. honey-bees) or true navigation (e.g. pigeons).

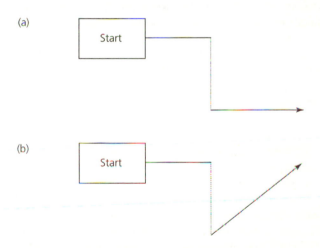

Figure 2.1 **The difference between compass orientation and true navigation**

(a) Compass orientation. After displacement (dotted line) the movement is in the same compass direction.
(b) True navigation. Adjustment is made to the course to compensate for the displacement. The animal arrives at its original goal.

Progress exercise

(a) What cues do you use when returning home from a local shop?
(b) If you were lost in a forest you would need different cues to find your way out. List some of the cues you might use.
(c) What types of orientation have you identified: pilotage, compass orientation or true navigation?

Use of landmarks

The use of landmarks in navigation is a familiar concept to humans. In the progress exercise many of you will have described the use of landmarks to find your way home. Towns and cities are, in effect, complex mazes and we use landmarks to find our way or to confirm that we are heading in the right direction. If we move to a new town or college we find it difficult to navigate. As we explore the area it gradually becomes familiar (Baker (1982) refers to this as a 'sense of location') and navigation becomes easier. It is probable that animals also learn about their local area. Pigeons become much more successful at homing after they are allowed to explore the region around their loft, and if repeatedly released from a distant location they show a marked improvement in returning – even in conditions designed to disrupt their homing. Pearce (1997) suggests that this reflects a growing reliance on local landmarks.

Research has shown that many animals use landmarks in homing behaviour. Typically this research has used the transformational approach. This involves training an animal to find a goal identified by experimental landmarks and then moving or changing the landmarks in some way. The animal's response to the transformation is then analysed (Cheng and Spetch, 1998). One of the earliest and perhaps the most famous example of this approach is Tinbergen's study (1951) of the digger wasp. This wasp digs a hole which it then stocks with its prey and Tinbergen was interested in how the wasp identified the location of the nest hole. He placed a number of pine cones in a circle around a nest hole and noted that on emerging from the nest hole the wasp flew around for a brief time before leaving to search for prey. He then moved the cones away from the nest and placed them in the same configuration. When the wasp returned it flew to the centre of the cone ring even though the nest hole was clearly visible a few centimetres away. The wasp seemed to rely on local landmarks to find its nest hole (Figure 2.2).

(a)

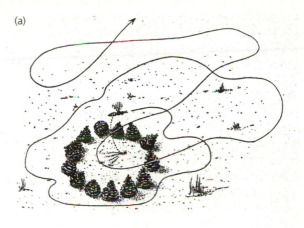

(b)

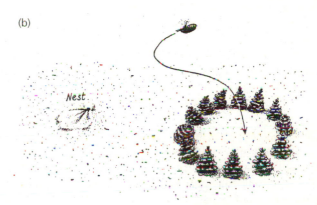

Figure 2.2 **The use of landmarks by the digger wasp**

    (a) The pine cones are placed around the nest and the wasp circles the area before flying off.

    (b) The pine cones are moved to a nearby location. On returning the wasp flies to the centre of the pine-cone circle.

Source: Based on N.Tinbergen, *The Study of Instinct*, copyright © 1951; reproduced by permission of Oxford University Press.

One problem of the Tinbergen study is that it does not identify which aspect of the circle of pine cones the wasp was using to navigate. It could have been responding to one particular cone (a single landmark or beacon) or to a number of the cones (multiple landmarks). A third alternative is that it was responding to the geometric relationship between a number of cones (configuration of the landmarks). A study by van Beusekom (1948) suggests that wasps use the information about configuration. Van Beusekom also placed a circle of pine cones around a nest hole. While the wasp was away he also constructed a square and an ellipse of cones next to the circle. The wasp never returned to the square, but returned to the circle and ellipse equally often. This suggests that the wasp used a mental representation of the circle of cones and was able to use it to distinguish between the circle and the square but not between the circle and the ellipse. There is now a great deal of evidence that nesting insects (ants, bees, wasps, etc.) use visual landmarks to navigate, sometimes over thousands of metres (Dyer, 1996).

Animals use a variety of information from landmarks to navigate: some use single landmarks, some use multiple landmarks and some use the configuration of landmarks. For example, Collet *et al.* (1986) found that gerbils are able to use one landmark to identify the location of food and can use information about the relative position of a number of landmarks. Cheng (1994) has found that pigeons use information about the relationship between a number of landmarks in their homing behaviour. In some species it is the geometric shape formed by landmarks, or configuration, that seems to be the important cue. For example, Cheng (1986) allowed rats to eat a small portion of some food in one corner of a rectangular arena. The food was then buried and the arena was randomly rotated before the rat was placed in the arena again. Cheng found that the rats searched equally in the correct corner and the one directly opposite. The tendency to search in the opposite corner has been called a 'rotational error' and this type of error is made despite a variety of visual and olfactory cues. Thus the rat may be shown food near a white wall smelling of peppermint but later searches by a black wall smelling of anise. It seems that it is the relationship between a short and long wall that the rat uses to initiate its search, and this relationship is the same in two diagonally opposite corners.

Olfactory cues

There are two ways odour can be used as a navigational aid in homing behaviour: odours can be used to identify the animal's home and act as an 'olfactory beacon'; or alternatively odours can be used to mark trails which the animal can follow to food sources or home.

There is evidence that the two species famous for their homing ability, salmon and pigeons, both use odours to aid navigation. In a series of experiments Hasler and his colleagues have shown that salmon identify their home stream using odour (e.g. Hasler *et al.*, 1978). When the noses of salmon are plugged, or the olfactory nerves are cut, the homing ability of the salmon is impaired. Similar experiments with pigeons have shown that they also need an intact olfactory sense to navigate to their loft (e.g. Wallraff, 1984). The problem with the experiments on both salmon and pigeons is that the procedure to block or destroy the sense of smell may cause side effects that affect their homing behaviour. It could, for example, affect the animal's ability to learn, or the stress of the procedure may make the animal vulnerable to illness or to predators (and if the animal dies it cannot home). An alternative approach is to manipulate the odours in the animal's environment and to observe the effects on behaviour. Grier and Burke (1992) describe both laboratory and field studies that seem to confirm that salmon use odours to find the appropriate waterway. In the laboratory study a number of salmon were raised in water containing different chemicals (odours). They were then tested in a four-armed maze, each of which contained water with a different odour that flowed into a central tank. When placed in the central tank the salmon swam into the arm that contained the odour in which they had been raised. In the field experiment some young salmon were exposed to one of two artificial odours (morpholine and phenylethyl alcohol) and were compared to a control group that were exposed to plain water. All the salmon were then released into Lake Michigan. When the salmon were due to return to the streams that feed into the lake one stream was scented with morpholine and another with phenylethyl alcohol. A further 17 streams along the same stretch of coast were monitored. The results were conclusive and showed a very high proportion of the fish exposed to morpholine returned to the stream with morpholine in. Similarly the salmon exposed to phenylethyl alcohol returned to the appropriate stream. The control group of salmon were recovered in a number of different streams. These results indicate that the homing behaviour of salmon is strongly influenced by the odours that they are exposed to when young.

There have also been a variety of studies of altering the olfactory cues for pigeons. One hypothesis is that pigeons home because they acquire an olfactory map of their local area by learning which odours are carried by different wind directions. If this were true then manipulation of odours and wind direction should disrupt a pigeon's navigation, and this seems to be the case. For example, Ioale *et al.* (1990) used air currents containing the odour of benzaldehyde in pigeon lofts. The air currents for the control group were in the normal direction, but for the experimental group fans were used to reverse the normal air flow. Both groups were exposed to the scent of benzaldehyde during transport to the release site. On release the control group set off in the correct direction but the experimental group went in the opposite direction. A number of studies have shown that pigeons' homing behaviour can be systematically disrupted by false olfactory information (e.g. Benvenuti and Wallraff, 1985).

However, the exact role of olfactory cues is unclear and there has been a failure to replicate some of the studies of such cues (Wallraff, 1990). Furthermore, pigeons raised in shelter from wind (and therefore receiving few olfactory cues) seem unaffected by procedures which block their olfactory senses (Able, 1996b). This suggests that olfactory cues may be one navigational cue used by pigeons but that it is not the only one. The theory that pigeons use olfactory cues to home has proved to be controversial and the debate about the exact role of odours in homing continues (e.g. Able, 1996b).

A variety of animals use odours to identify routes to food sources and then back to the nest or home. For example, many species of ant lay down a **pheromone** trail between a food source and their nest (Pearce, 1997). Pheromones are chemicals that are used to transmit information. Each ant adds to the pheromone trail as it travels back to the nest until the food source is gone; then they stop and the trail dissipates. Some species of loris lay down scent trails around their home range by using a behaviour known as urine washing (Shorey, 1976). This involves urinating onto their hands and rubbing their hands and feet together before setting off through the branches of trees. As they grasp branches and twigs they leave a scent of urine; this can then be retraced and used to find home territory even in complete darkness.

Retracing routes and 'dead reckoning'

One obvious way of returning back home is to retrace the route you have taken on your outward journey. If you go 100 paces north and then 50 paces east you can return home by going 50 paces west then 100 paces south. There is evidence that animals learn routes in their home territories. A famous illustration of this is provided by the description of 'path-habits' in water shrews by Lorenz (1952). He described how the water shrews traced routes and would jump over obstacles and on and off stones etc. However, when the stones were removed the water shrews would still jump at that spot and land with a bump! A similar phenomenon has been observed in bats returning to their caves. The bats often collide into objects introduced to the cave even though they should detect them (Griffin, 1958).

However, both these examples illustrate what Shettleworth (1998) calls 'response learning' rather than place or direction learning. In other words the animals had learned a series of responses such as turn left, run 20 centimetres, jump onto stone, etc. Thus when the environment was changed the animal made mistakes because it followed the same set of responses. However, animals also seem capable of direction learning and there is some evidence that their homing behaviour can be influenced by experience of the outward route. For example, when two groups of pigeons were taken from the same loft to the same release point by two different routes they took different initial headings home (Baker, 1980). Each group set off in the approximate direction that they had brought to the release site, which suggested that they had learned something about the direction of the outward route. However, even if pigeons are anaesthetised during their outward journey to the release site (and therefore unable to learn about the route) they are still able to home (Walcott and Schmidt-Moenig, 1973).

Retracing an outward route home is often an inefficient means of homing since the outward route may involve many twists and turns. A direct route home is often much quicker. If, for example, I go north 100 metres, then east 100 metres and then south 100 metres I would have to travel 300 metres to return 'home' if I retraced my steps. However, if I were to head west I would be back home in only 100 metres (Figure 2.3). Assuming there are no landmarks the only way to do this is to have some record of the distances and directions travelled from the home point. This type of navigation is called **dead reckoning** and, although it would seem to use complex computations, there is evidence that animals can use it in homing behaviour. For example, **foraging** desert ants search for food in a seemingly haphazard way, zigzagging away from the nest. The search may take them over a 100 metres from the nest (although the ant will have walked much more than a 100 metres to get there). Nevertheless, as soon as the ant finds prey it carries it in a straight line back to the nest. If, however, the ant is picked up as soon as it discovers prey and displaced several 100 metres, it now heads for where the nest would have been if it hadn't been moved (i.e. it uses compass orientation). Furthermore, when it arrives at the point where the nest should have been it begins a search for the nest, no longer travelling in a straight line but moving in a spiral (Wehner, 1992). Of course, there is the possibility that, even though the ground in a desert is fairly featureless, the ants use some form of landmark to identify the nest. However, if ants are trapped as soon as they emerge from the nest and moved 5 metres away they do not seem able to return to the nest. It seems unlikely that the ants can use a landmark from a 100 metres but not 5, therefore the ants seem to be using dead reckoning to return. Honeybees also seem to use dead reckoning both to return to the hive and to food sources. Furthermore, honeybees communicate information about the direction and distance of food sources to other bees, which then fly directly to the food source (Von Frisch, 1950).

There is evidence that a variety of animals are capable of dead reckoning, including mammals and birds. For example, if a gerbil pup is removed from its nest the mother begins to search for it. The search usually involves a zigzagging path before the pup is found. However, when the mother finds the pup she takes it in a straight line back to the nest even in total darkness when no visual cues are available (Mittelstaedt and Mittelstaedt, 1982). There is also evidence that animals can use dead reckoning even if they have been passively transported. Saint Paul (1982) took seven geese on an open cart to a release point via a semi-circular route. When they were released they walked directly towards home. A similar group which had travelled in a covered cart did not head home. This suggests the first group had learned about the outward route.

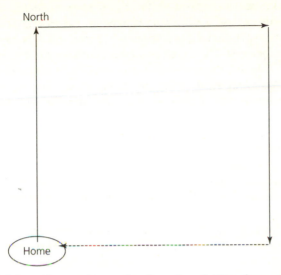

Figure 2.3 **Retracing routes can involve a long journey home. There are often shorter alternatives**

Can humans use dead reckoning? Try testing a friend.

You will need a blindfold and a large area to test your friend's ability to use dead reckoning (a hall or a field). Start by marking a homing point and then lead your blindfolded friend by a haphazard route away from it. Then ask your friend to return directly to the homing point. Try the same process with a spiral route away from the homing point.

Did you find evidence of dead reckoning? What other cues might have been used to find the homing point?

Progress exercise

Compass navigation

One hypothesis of how pigeons are able to home is that they possess a map of some sort and this allows them to determine which direction to fly (Kramer, 1952). This hypothesis can only work if the pigeons have some way of orienting themselves to find the correct direction; in other words they need a compass of some kind. There are a number of features in the environment that could act as a compass for pigeons and other animals: the earth's magnetic field, the sun and the stars.

Magnetic compass

Humans have long used a magnetic compass to navigate by using devices that point to the magnetic poles and indicate the direction of north and south. It is possible that animals also use information about the earth's magnetic field to navigate, but to do so animals must be sensitive to magnetic fields and be able to use this information. There are two types of study of magnetic compass in animals: one designed to test the animal's sensitivity to magnetic fields and the other designed to disrupt the magnetic information in order to find out if this disrupts navigation.

Many of the early studies of a magnetic compass tried to disrupt the homing ability of pigeons by placing magnets or coils that generate magnetic fields to distort magnetic information. Initially these studies failed to show any disruption to the pigeons' homing ability, either because pigeons do not use a magnetic compass or because they were using another compass to compensate. Keeton (1974) argued that it was the latter and proposed that pigeons navigate using either a sun or magnetic compass. He found that if pigeons carrying magnets (which disrupt their ability to detect the earth's magnetic fields) are released on overcast days their homing is impaired. However, Papi *et al.* (1992) have argued that this is because of the side effects which the magnetic fields have on the nervous system rather than to a magnetic compass. They found that changes to magnetic fields affected the opioid system in pigeon brains (this system controls pain responses). Furthermore pigeons that were given a drug that blocked the action of the opioid system also showed problems in homing even though they were not carrying magnets.

Attempts to demonstrate that pigeons can detect magnetic fields have produced mixed results. Some studies have used classical conditioning to test pigeons' responses to magnetic fields. In these studies changes in a magnetic field have been paired with a mild electric shock which triggers an increase in heart rate. If the pigeons were able to detect the changes in the magnetic field then at the end of the study changes in magnetic field alone should have also caused an increase in heart rate. However, in most studies there was no change in heart rate and a consequent failure to demonstrate sensitivity to magnetism (e.g. Kreithen and Keeton, 1974). This is a problem for the magnetic compass theory, if an animal cannot detect something how can it respond to it? One study, though, has found that if pigeons are allowed to move rather than being held immobile in experiments then they could detect magnetic fields (Bookman, 1978). It is possible that pigeons are sensitive to movement in a magnetic field rather than to the magnetic field itself. In a review of magnetic orientation in birds Wiltschko and Wiltschko (1996) conclude that a magnetic compass has been demonstrated in both homing and migration. They suggest that the evidence of magnetic orientation points to an 'inclination compass'. Magnetic field lines intersect the surface of the earth at different angles according to latitude (Figure 2.4). At magnetic north and south the fields intersect the surface at a perpendicular angle, but at the magnetic equator the fields run parallel to the surface. Thus the angle that the lines enter the surface indicates 'poleward' or 'equatorward'. Wiltschko and Wiltschko (1996) also note that magnetic parameters, such as local distortions in the magnetic field, can act as a magnetic map which birds can use to determine home direction.

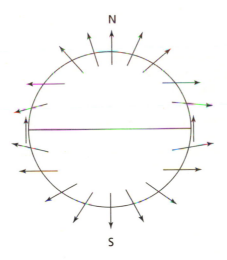

Figure 2.4 **A map of the earth's magnetic fields**

Sun compass

There is a variety of evidence that suggests that animals can and do use the sun to navigate. For example, pigeons can be trained to use the direction of the sun to locate food, and if the apparent direction of the sun is distorted using mirrors the pigeons shift their search for food accordingly (Kramer, 1952). However, there is one major problem in using the sun as a compass direction: it moves. To be able to use the sun as a navigational aid animals must have an internal clock to compute the direction of the sun. Imagine an animal needs to head south-west. As the sun is only in that direction in mid-afternoon, at any other time of the day the animal has to extrapolate from where the sun is to where it would be mid-afternoon. Animals' ability to do this has been tested using experiments involving **clock-shifting**. It is assumed that any internal clock is regulated by the light–dark cycle and that it can be manipulated by artificially shifting the cycle. Humans experience this shift in the internal clock when they travel any distance east or west. If animals are kept in an environment where external cues are excluded it is possible gradually to shift their internal clock so that the animal is ahead or behind real time. Thus the lights may go off in the artificial environment indicating 'dusk' but in the real environment it is really midday. If an animal has had its internal clock shifted then any directional cues taken from the sun's position will be wrong. For example, if a pigeon has had its internal clock shifted ahead six hours and is released at mid-morning the sun will *actually* be in the south-east but the pigeon's clock is set at mid-afternoon and it will *infer* that the sun is in the south-west. Thus if pigeons use a combination of the sun's position and an internal clock they should head at right angles to the correct direction. A number of experiments have confirmed that this is the case in pigeons (Keeton, 1969) and honey-bees (Renner, 1960). Pigeons appear to be able to use the sun compass for true navigation and can use it to return to their loft no matter where they are released. Honey-bees seem to use the sun for compass orienting (e.g. a cue for direction) since if they are displaced they have problems finding locations (see Chapter 7, pp. 113–116, for a full discussion).

Although the sun appears to be an important source of information about direction for animals there are problems connected with relying on it. One problem is that it cannot be used on very overcast days, thus if it were the sole navigational aid then animals would be unable to navigate in heavy cloud. However, a number of animals seem to use several types of navigation. For example, clock-shifting only affects the ability of pigeons to home on relatively sunny days, and on very overcast days clock-shifted pigeons home equally as well as control pigeons (Keeton, 1969). This suggests that pigeons use a sun compass when the sun is visible (hence disruption to clock-shifted birds) but use a different compass, possibly magnetic, when it is overcast (clock-shifting does not affect magnetic fields). Another problem of using a sun compass is that it cannot be used at night and, although some animals rest at night, others are on the move.

Star compass

Most of the studies of a star compass have concentrated on its use in migration but it is included here as another compass technique. At night stars form a changing pattern that can be used for information about direction and time of year. There is evidence that some animals may use this information to navigate. For example, buntings that were reared without sight of the sky were unable to navigate during migration (Emlen, 1972). In another study Emlen (1975) exposed young buntings to artificial night skies in a planetarium and, using the direction of migratory restlessness (see p. 34) found that they learned about the position of stars and orientated themselves by using the position of the pole star relative to the other stars.

Conclusion

There is evidence that animals can use magnetic compass, sun compass and star compass to navigate and home. Compasses are useful when the type of navigation required is compass orientation. However, a compass only indicates a direction and is only useful if the animal knows where it is in relation to where it is going. Therefore for true navigation, such as when a pigeon is released from a random point in the country, a compass only works with reference to a map (and is called compass and map navigation). Pearce (1997) points out that the evidence for a map component is weak (see 'Cognitive maps', in Chapter 6, this volume) and he therefore concludes: 'As it stands, then, the map and compass hypothesis must be regarded as an incomplete account of homing' (p. 219). Nevertheless pigeons (and other animals) are able to navigate home from unknown, distant release points. Even if the map and compass hypothesis is an incomplete account, it seems the most plausible explanation of this ability.

Summary

Animals use a variety of types of navigation in homing and migration, including pilotage (using features from the environment to orientate), compass orientation (heading in one direction) and true navigation (finding a goal no matter where the start point is located). The use of visual landmarks is a form of pilotage and is a common method of homing. Animals may use single or multiple landmarks or the configuration of landmarks in homing. Olfactory cues can be used as a cue for pilotage and are used for homing behaviour by a variety of animals. Olfactory cues can also be used to lay trails to and from home that can be followed by animals. Animals can learn routes that can be followed in their home environment. However, it is not always clear whether the animals show response learning or whether they learn about the place. Dead reckoning is a method of homing that requires a calculation of the distances and directions that have been travelled since leaving home. Then, no matter how complicated the route out, it should be possible to return directly home. Despite its seeming complexity some animals are capable of dead reckoning. Another way of homing is to use some form of compass to navigate, and it seems that the earth's magnetic field, the sun and the stars can all be used as a navigational aid. However, compasses are only useful for true navigation if the animal possesses some form of map but the evidence for this component is weak.

Review exercise

Briefly describe the following concepts in your own words:

1 transformational approach
2 dead reckoning
3 sun compass
4 magnetic compass

What is the problem of using compasses for true navigation?

Further reading

McFarland, D. (1999) *Animal Behaviour* (3rd edn), Harlow: Longman. There is a good section on navigation in the chapter on 'Co-ordination and Orientation'.

Pearce, J.M. (1997) *Animal Learning and Cognition* (2nd edn), Hove: Psychology Press. This book contains a good discussion on navigational methods and homing which is presented in a very readable style.

3

Animal navigation: migration

Introduction

Many of the navigational techniques discussed in the previous chapter on homing behaviour are also relevant to migration. In long-distance migration animals also use techniques involving pilotage, compass orientation and true navigation (see pp. 16–17 for a description of these types of orientation). However, in migration everything is typically on a much larger scale. For example, instead of using a tree or a bush as a landmark for a nest, birds may follow the coastline of a continent or use major rivers to guide their migratory routes. Also, the distances involved tend to be much greater, sometimes thousands of kilometres instead of a few metres or several kilometres. The Arctic tern, for example, flies from the Arctic to the Antarctic and back again each year – an annual journey of 35,000 kilometres.

There are a number of types of migration, including dispersal and seasonal migration. Dispersal migration occurs when animals spread out to explore and colonise new habitats and involves the search for favourable ecological conditions. The type of migration being considered in this chapter is seasonal migration, which typically involves mass movements away from and back to breeding and feeding areas. Able (1980) defines migration as the 'oriented, long-distance, seasonal movement of individuals'. There are two main questions about the cognitive aspects of migration: how do animals navigate during migration and what initiates migration? A third question, which is related to both cognitive and evolutionary processes, is 'why do animals migrate?'

How is migration studied?

Many species of animals migrate and because of the diversity of their habitats a number of specialist techniques are remained to study migration. However, there are some common methods of investigating migration. These include:

1 *Tracking*. There are numerous ways of tracking the migratory patterns of animals. One is simply to follow the animals on their migratory path. However, it is difficult for humans physically to follow birds or aquatic animals and in these circumstances tagging would be preferred. Animals can be tagged at one point (perhaps at the breeding grounds) and then the tagged individuals can be identified at other points en route or at various destinations. Electronic tagging allows the tracking of migration at a distance, and recent advances in satellite tracking allow the path of individual animals to be mapped accurately over great distances (e.g. Papi and Luschi, 1996). Tracking can show what happens during migration and where the animals go, but it does not show *how* the animal navigates or *what triggered* the migration.

2 *Behaviour of caged animals*. Animals, particularly birds, often show heightened activity if they are caged during the time they would normally be migrating. This '**migratory restlessness**' can be used to test what triggers the migratory behaviour. Various environmental cues (such as temperature, day length, etc.) can be manipulated to find which, if any, cause the migratory restlessness. Also, caged birds in circular cages tend to jump and agitate in the direction that they would be migrating and this tendency can be recorded using

inkpads in the cage. Thus there is a record of both the time and the direction of migratory restlessness.

3 *Transport or displace animals*. There are two types of information that can be gained from displacing a migrating animal. One is the direction an animal takes on release from the site it is transported to and the other is whether transporting the animal affects its eventual destination. Both types of information help determine whether the animal is using compass orientation or true navigation to guide migration.

4 *Impair senses*. This technique was discussed in Chapter 2 in relation to olfactory cues in homing behaviour but it is a useful technique in studying migration. Senses can be impaired either by physically blocking the sense, by impairing the sensory ability surgically, or by distorting the sensory input. This allows investigation of which senses are used in navigation.

5 *Clock-shifting*. Clock-shifting was discussed in the previous chapter in relation to homing using the sun as a compass. It is also a very useful technique to study the use of compass navigation in migration.

Examples of navigation techniques in migration

Migration occurs in a number of different types of animal and the navigation techniques they use vary considerably from species to species. However, the kind of navigation used can depend as much on the mode of travel as on the type of species. Navigation when walking through mountain passes requires different techniques to navigation across vast expanses of ocean, and crossing an ocean underwater demands different techniques than flying over it. One common theme that emerges from the study of animal migration is that many animals use a number of different cues to navigate. For example, the orientation of loggerhead sea turtle hatchlings seems to be a complex inter-action of visual information, wave direction and a magnetic compass (Lohmann and Lohmann, 1996). Able (1996b) points out that 'animal orientation systems are replete with interacting mechanisms and are highly flexible' (p. 1), and he warns that understanding of orientation will only be achieved by acknowledging its complexity.

Migration using flight

Flying allows animals to cross any type of surface (land, marsh or water) with equal ease and a number of flying animals migrate. The most widely studied group is birds, possibly because of the numbers involved and the distances travelled, but bats and insects also use flight to migrate. The distances covered and the accuracy of these aerial migrations suggest that the animals must use one or more navigational techniques. The arctic tern, for example, breeds in a number of areas in the Arctic Circle in summer and migrates to the Antarctic in winter. This involves a round trip of between 30,000 and 40,000 kilometres, but they return to the same breeding grounds year after year. Birds are capable of precise navigation, ringed birds having been found to return to highly specific locations (such as one burrow on an island) for many years in a row (Waterman, 1989).

There have been many studies of how migrating birds navigate, but, as Grier and Burke (1992) note, much of the evidence has caused confusion. Part of the confusion has arisen because of the diversity of mechanisms used by birds to navigate but more problems have been caused by the fact that many species seem to have redundancy built in to their navigational systems. That is, like the homing pigeon, many migrating birds use more than one means of navigating and may use different systems according to the circumstances. Furthermore, even within the same species there is variation in the navigational system used depending on experience and environment. For example, Perdeck (1958) captured and ringed juvenile and adult starlings that were en route from their breeding grounds in the Baltic to the wintering grounds in southern Britain and northern France. The birds were captured in Holland but were released in Switzerland; this displaced them 750 kilometres to the south-east (Figure 3.1). When the final destination was analysed the juvenile and adult birds tended to be in different locations. The juvenile birds were found in southern France and northern Spain. They seemed to have used compass orientation and had carried on flying in the same direction as if they had not been displaced. However, the adult birds were found in their normal wintering grounds. They had compensated for the displacement and showed evidence of true navigation, presumably because of their experience of previous migrations.

There is evidence that birds use a variety of compasses. Clock-shifting experiments show that diurnal migrants such as starlings can

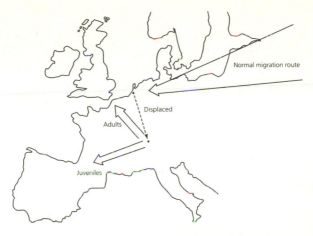

Figure 3.1 **The result of displacing starlings during migration**

The juveniles resumed the same compass heading and went south-west to northern Spain. The adults corrected for the displacement and headed north-west to their normal wintering grounds.

Source: Based on Perdeck (1958); reproduced with permission of Ardea.

use the sun as a compass. For some nocturnal migrants the position of the sun as it sets is important in orienting themselves, but for others the stars act as a compass. For example, mallards released on a clear night all depart in a consistent predicted direction. However, if released on an overcast night they depart randomly (Bellrose, 1958). Many species of bird seem to use the sun or stars as a compass but are still able to navigate on very overcast days; it seems that these species are also able to use a magnetic compass. Grier and Burke (1992) point out that the magnetic sense is not well understood but that it is 'clear that it exists and it influences the orientation of many species of bird' (p. 228). More recently Wiltschko and Wiltschko (1996) have demonstrated that a number of birds use the inclination of magnetic fields to orientate (see p. 27, this volume). In birds that stay in either the northern or southern hemisphere this information is adequate to indicate 'poleward' or 'equatorward'. However, long-distance migrants that cross the equator face a problem since the information about the inclination of the magnetic field becomes reversed. Wiltschko and Wiltschko (1996) suggest that these long-distance migrants use information about the magnetic inclination and celestial cues, and that the two systems interact in a complex way. Satellite tracking of albatrosses shows that they are able to pinpoint small islands by following straight lines which compensate for wind direction even after long-distance displacements (Papi and Luschi, 1996). There is evidence that the albatrosses use sun, star and magnetic compasses, but Papi and Luschi believe that the position-fixing capacity of the birds cannot be fully explained by known navigational mechanisms. It is possible that there are more navigational techniques yet to be uncovered.

Box 3.1 The monarch butterfly

The monarch butterfly is a spectacular aerial migrant. Each autumn millions of monarch butterflies migrate from southern Canada and north-eastern USA to fir forests in the mountains of central Mexico. They remain there for the winter, congregating in enormous numbers on the fir trees before returning north in the spring. Marked individuals have been found to travel 80 kilometres a day, and the round trip may be as much as 8,000 kilometres. The locations of the fir forests in Mexico are very specific yet the butterflies are able to navigate to them. What makes this remarkable is that the journey south is made by novices since the adults returning to the north lay eggs then die. The butterflies born in the summer have a short lifespan of 4–6 weeks. Thus the butterflies flying south in the autumn have never done so before and have no experienced individuals to follow; yet they arrive at the same small fir forests used by their ancestors. Experiments conducted by Brower (1996) in Kansas suggest that the butterflies use a sun compass and endogenous clock. He took some butterflies from trees at night and then at 1 a.m. he put on a bright light to imitate sunrise. The normal hour for sunrise was at 7 a.m. When released the clock-shifted butterflies flew at 90 degrees to the correct position.

Anyone who would like to follow the monarch migration in spring or autumn should visit the monarch watch website (www.monarchwatch.org) which contains a variety of information about monarch butterflies, including the annual progress of the migration.

Aquatic migration

More than 70 per cent of the earth's surface is covered in ocean and, despite the difficulties in studying marine life, it is known that a number of animals migrate over great distances. In addition some animals migrate in large lakes and in large river systems. The techniques used in underwater navigation tend to differ from those used by flying animals because of one major factor: visibility. Even under *optimal* conditions visibility underwater is poor and light does not penetrate more than about 1,000 metres. Thus any use of landmarks for piloting is limited to coastal regions and clear rivers or lakes. For example, grey whales migrate from the Arctic Circle to an area off Mexico to breed by following the east coast of North America, and they seem to travel in straight lines from one promontory to the next. Whales have an advantage over many marine animals because they can swim on the surface and are not restricted to cues that are only visible underwater. The poor visibility underwater restricts the opportunity to use celestial navigation since the sun and the moon can only be seen 10 metres or so from the surface (again in optimal conditions) and stars can only be used to navigate very close to the surface. Nevertheless some fish are known to use a sun compass and it is likely that others use the polarisation of light as a navigational cue (Braithewaite, 1998). Aquatic animals have evolved a number of senses that do not require light such as echolocation and electric location, but little is known about their role in migration.

One cue that is available during migration in oceans is the current. The currents in the oceans in the northern hemisphere tend to be clockwise whereas in the southern hemisphere they are anti-clockwise. There is evidence that animals may use these currents to guide them on long-distance migrations. For example, the North Atlantic bluefin tuna breed in the Gulf of Mexico and migrate to the coast off Scandinavia to feed following the Gulf Stream (Waterman, 1989). Loggerhead turtles which hatch on the beaches of Florida seem to follow the clockwise currents of the North Atlantic to swim in loops around the Sargasso Sea before returning to the same beach in Florida (Pearce, 1997). It is not known how the turtle returns to the beach it hatched on but it has been suggested that they might use chemical cues, ocean swells or magnetic fields. However, none of these theories have received much support. Satellite tracking of green turtles shows that they can navigate precisely back to an island even after displacement, and it is suggested that they use a variety of cues (Papi and Lusci, 1996). Salmon are known to use olfactory cues to identify the stream in which they spawned and use this information to guide their migration to breed (Hasler *et al*., 1978).

Migration on land

A number of animals migrate on land, principally large mammals such as wildebeest, bison, and elk. The movement of these animals can be followed either on the ground or via satellite tracking of radio collars. The distances travelled in land migrations are typically less than those using flight or even aquatic migration because the energy requirements of land travel tend to be greater. Land migrations are also often limited by physical barriers such as mountain ranges. Animals migrating on land tend to use pilotage to navigate since over the relatively short distances animals can use landmarks such as mountains and rivers (Waterman, 1989). Many of the migrating mammals travel in large herds consisting of a number of generations, and it is possible that the young learn the routes and landmarks from the older generations. A spectacular example of this is the annual mass migration of some 1.4 million wildebeest through Tanzania and Kenya. The wildebeest spend most of their lives on the move following the seasonal rains (and thus the best feeding conditions). Wildebeest calves are born in the south of the Serengeti Plain in Tanzania in February, and in March the wildebeest begin to head north and west towards Lake Victoria (Figure 3.2). The breeding season is in June in the north of Tanzania and in August they cross the Mara River to the Masai Mara in Kenya. In November the wildebeest start to head south towards the Serengeti Plains where the cycle begins again in February. Thus the life of a wildebeest is an

Progress exercise

Long before the advent of magnetic compasses, clocks and maps Polynesian sailors made round trips covering vast distances between islands in the Pacific Ocean. Review the navigational techniques mentioned above and list some of the ways the sailors might have used them. Can you think of any additional cues that might be used?

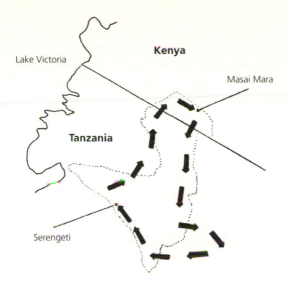

Figure 3.2 **The migration route of wildebeest**

almost constant clockwise migration around the plains of East Africa, with many generations travelling in a mass herd.

Migration: endogenous or environmental control?

Seasonal migration, when animals move from one habitat to another at seemingly fixed times, raises the question of what causes this behaviour and what cognitive processes are involved. Of course, because of the diversity of animals there is no one answer to these questions but there are a number of possibilities:

- migration may be initiated by **endogenous factors** (by an internal clock for example);
- migration may be triggered by a combination of endogenous and environmental factors;
- the timing and route of migration may be learned by young animals as they follow adults.

There is evidence that both the timing and direction of migration have some form of endogenous control in some species. For example, the garden warbler and the willow warbler spend summer in Europe and winter in Africa. They show marked **circannual** (approximately yearly) changes in body weight, moult and food preferences that are linked to their migration. Birds which have been raised in constant laboratory conditions show the same seasonal changes as free-living birds, suggesting that the changes are under internal control (although the timing of the changes was less than a year, which may mean that environmental factors 'set' the circannual rhythm).

Gwinner (1986) believes that migration is initiated in many birds by an internal circannual rhythm rather than by environmental changes. Also, by comparing the behaviour of wild birds with the migratory restlessness of caged birds, Gwinner has found evidence that birds have an endogenous clock which induces them to fly for as many hours as are needed to reach their destination. Captive birds showed migratory restlessness for the length of time it would take to reach their normal destination. There is also evidence of an interaction between **circadian** (approximately daily) and circannual rhythms in bird migratory behaviour. For example, birds that are normally active during the day but migrate nocturnally seem to have circannual signals that cause alterations in circadian activity, leading to an increase in nocturnal activity (Gwinner, 1996). There is also evidence that the direction of migration may be under genetic influence. For example, white storks breed in Europe and over-winter in Africa, but different populations take different routes. Those from the west cross the Mediterranean via Gibraltar (they head south-west), whilst those from the east go via the east coast of the Mediterranean (they head south-east). However, if a young stork is taken from the east and released in the west it heads south-east, suggesting that the direction taken was innate.

There is also evidence that environmental factors play a part in the timing of migration. The changes in day length that accompany the seasons in high northern and southern latitudes act as a reliable indicator of the onset of winter or summer. A number of migratory birds use this information to initiate migration (Ridley, 1995). When these birds are put in an artificial environment where the day length is shortened they show migratory restlessness, suggesting they react to day length not the amount of time that has elapsed. Temperature and the availability of food can hasten or delay departure beyond those normally expected in a usual season, which suggests that there is an interaction between endogenous and environmental factors (Waterman, 1989).

In some species environmental conditions may induce or cancel migration. For example, lapwings from some areas of western Europe only migrate south in particularly harsh winters.

Some species of mammal which travel in large herds may learn migratory routes and the cues that trigger migration as the young accompany older generations (Grier and Burke, 1992). For example, if young bighorn sheep are artificially translocated they cease to migrate, presumably because they have not learned where or when to travel. Also, if traditional routes are blocked adult bighorn sheep do not take alternative routes.

It seems that for most animals the trigger to migrate is a complex interaction of factors and that a variety of information needs to be processed. In his conclusions about migration Waterman (1989) notes: 'the total sensory, learned and genetic information available to the animal must be highly "distilled" to allow the best choices for action' (p. 224). The exact nature of this information-processing and consequent actions will vary from species to species. For example, the information that initiates the migration in a humpback whale is unlikely to be the same as the information that triggers migration in a monarch butterfly.

Arctic terns breed as far north as the Arctic Circle in the summer and spend winter feeding as far south as the Antarctic Circle. They can cover thousands of kilometres in a few days but when they need to rest they do not have to find land since they can rest floating at sea. Imagine an arctic tern that feeds south of Australia during the winter and breeds in the Orkney Islands (off the north-east coast of Scotland) in the summer. Write a brief description of some of the navigational cues the bird might use to return to its breeding grounds. What cues might trigger its return to the south?

Progress exercise

Why migrate?

After studying the navigational methods that animals use to migrate, and the cues that initiate migration, there remains one fundamental question: why do animals migrate? The standard answer is that, on average, the benefits of migration outweigh the costs. There are many costs involved in migration but the most obvious is the energy expended on the journey. Some animals travel weeks or months during a year, often with limited opportunities to feed; therefore their food reserves become severely depleted. Many species of whale feed very little on their way from polar feeding regions to the temperate breeding areas. Another cost lies in the potential dangers to the animal en route. Wildebeest follow the seasonal rains and have to cross fast-flowing rivers, with a number dying from drowning each year. Birds crossing mountains or seas risk freezing or drowning. Finally there are 'biological costs' such as the development of navigational systems or moulting twice a year to prepare for migratory flights.

Since the costs of migration can be very high one might expect there to be great benefits. Migration involves moving to an area where the conditions for feeding or breeding are better than if the animal was sedentary. Alcock (1993) points out that migration occurs in many species but that 'one convergent pattern of migration involves species whose members move to an area when that region is rich in resources and then leave it when the place becomes less suitable' (p. 293). For example, high northerly or southerly latitudes have marked seasonal changes with cold, short winter days and warm, long summer days. However, the seasons are opposite – summer in one hemisphere is winter in the other. A number of birds take advantage of this and move north and south to the warm conditions and longer days to optimise the potential food source. Not all migration involves maximising food availability however – some animals migrate away from rich feeding grounds to reproduce. For example, many marine mammals leave their feeding grounds to find isolated regions to rear young (e.g. seals) or warmer seas to give birth (e.g. whales).

There are two types of evidence for the cost–benefit hypothesis. Firstly, the cost–benefit hypothesis may explain why in some species (principally birds) individuals may migrate one season but not another. It is more difficult to survive in harsh winters and the benefits of migration are great, but in a mild winter the costs of migration may outweigh the benefits. Berthold (1998) cites a number of changes in avian migratory behaviour that may be a reaction to global warming. These include a reduction in migratoriness, later departures, earlier returns and a reduction in migratory distance. Secondly, the cost–benefit hypothesis could explain why some species do migrate but others do not since the cost and benefits will differ. For example, many of the birds that migrate south for the winter from northern Europe are insect eaters but those that remain are seed eaters. The cost of migration would be approximately the same for all birds of a similar size, but the benefits for insect eaters are enormous: there is abundant food in the south but it is very scarce in the cold north. However, the benefits for the seed eaters would be less marked since there are still seeds available in the north in winter.

There are doubts whether the cost–benefit hypothesis can explain all migratory behaviour. Waterman (1989), for example, notes that the costs of migration sometimes seem disproportionate to the benefits.

Summary

Migration typically involves long-distance seasonal movements to and from breeding and feeding areas. It is a widespread phenomenon and is seen in birds, fish, mammals, insects and amphibians. Birds can cover vast distances and often fly across or between continents. However, the use of flight in migration is not confined to birds; it is also used by a number of insects. Animals can also travel long distances in aquatic migration, but despite the difficulties navigating in oceans they seem to find their destinations with precision. Animals use a wide variety of techniques to navigate during migration and a number of species, particularly birds, use several navigation systems. These include pilotage, using landmarks and a variety of compass techniques such as sun, star and magnetic compass. The control of when and where to migrate can be endogenous or learned depending on the species, but in most it appears to be a complex interaction of endogenous factors, learned factors and environmental cues. The reasons why animals migrate are also diverse but it seems that in general terms the benefits of migration outweigh the costs.

One of the most spectacular sights in nature is the migration of vast herds of wildebeest through Tanzania and Kenya. A puzzle of nature is how turtles can migrate to the precise beach on which they hatched after spending years swimming around an ocean. Write a brief account of the different navigational cues used by these two forms of migration.

Review exercise

Further reading

Grier, J.W. and Burke, T. (1992) *Biology of Animal Behaviour*, Dubuque, Ia.: Wm. C. Brown Communications. This book has chapters covering most aspects of animal migration and is written in a clear and interesting fashion.

Waterman, T.H. (1989) *Animal Navigation*, New York: Scientific American Press. This book is devoted entirely to the subject of animal migration and navigation. It is easy to read and understand and is an excellent text for anyone with a strong interest in this area.

The January 1996 issue of the *Journal of Experimental Biology* was devoted entirely to the topic of navigation. The articles cover most of the debates and issues relating to navigation and migration and are written by some of the best-known authorities on the subject. The articles are not aimed at A-level students, but they are interesting and illustrate some of the current debates.

Animal communication

Introduction

My partner and I have spent a number of summers travelling in south-east Asia. One year while walking along a beach in west Malaysia we saw a troupe of monkeys in a tree and stopped to look at them. They soon became very agitated and several large males advanced on us baring their teeth; we beat a hasty retreat. A year later we were in a reserve in the south of Java and a similar incident occurred: we had to walk backwards away from the monkeys waving sticks to prevent an attack. When the same thing happened yet again in the hills of Bali we began to suspect that one of us was somehow triggering these attacks. When we saw a television programme that showed some of the threat signals used by monkeys it dawned on us that perhaps I caused them. One way some monkey species threaten each other is to use an 'eye-lid flash' which involves raising the eyebrows to flash lighter patches of fur. Another way of showing aggression in many species of monkey is a prolonged stare. I have large dark eyebrows that I raise frequently and, when something interests me, I look at things intently. It is possible that, like Dr Dolittle, I can communicate with monkeys, but all that I communicate is aggression!

This example raises some of the central questions to be discussed in this chapter:

- What is communication?
- How can we tell if communication has taken place?
- How do animals communicate?

The study of communication is central to the study of animal behaviour since it is the means by which animals interact with each other. All animals, even the most solitary, interact with other animals to some degree and many live in social groups. Dawkins (1995) points out that 'communication is the fabric of animal social life' (p. 71) and that most aspects of an animal's life (courtship, mating, parenting, fighting, etc.) require one animal to influence another. Given the importance of animal communication it is not surprising that there have been numerous studies of it. However, the picture that emerges from the research is confused, largely because of problems of definitions and measurement. Although the meaning of the term 'communication' seems clear and simple the study of animal communication has been dogged by the lack of a universally accepted definition. It is therefore necessary to start this chapter by examining some of the ways that the term 'communication' has been interpreted. The concept of language, which is a special and complex form of communication, is discussed in the next chapter.

Definitions of communication

The common-sense view of communication is that it is the transmission of information between two or more individuals through the use of some type of **signal**. In order to determine whether communication has taken place we can observe whether one animal's behaviour has an influence on the behaviour of another. Unfortunately, the complexities and diversity of animal communication has led to decades of debate about the exact nature of communication, and the terms 'information', 'signal' and 'influence' have been subject to intense debate.

Some of the problems of defining communication are best illustrated by considering some examples of animal behaviour. A male blackbird singing from the top of a bush is sending information about territory and, if this deters other male blackbirds from approaching, it has influenced the behaviour of others. This seems to correspond with the common-sense view of communication. However, in advertising its presence to other blackbirds the male blackbird also advertises its presence to the neighbourhood cats and may well influence their behaviour: they could begin to stalk the bird. Is this an example of communication? The example fits the description of communication given above since the blackbird has sent information about its position to the cat and this information has influenced the cat's behaviour. However, this does not take into account the purpose of the singing. Some researchers take the view that communication has evolved to confer some evolutionary benefits to the sender and that true communication involves specially evolved signals or displays (e.g. Krebs and Davies, 1993). In the example above the blackbird song has evolved to signal territorial claims and benefits the blackbird by discouraging other birds from encroaching; it did not evolve to encourage attacks from predators. This does not mean that communication always has to be between animals of the same species as a number of animals use signals that have evolved to send information to other species (to deter or confuse predators for example).

Another source of confusion has been the term 'information', which has been used in two different ways. The word 'information' has been used by some to refer to content or meaning. This is the sense of the term 'information' when we communicate using language. If I tell you that you have a smudge of blue ink on your left cheek then you understand the content of the message. You *know* what you look like, where the ink is and what colour it is. However, the assumption that information conveys meaning to animals is a subtle form of anthropomorphism since we cannot be sure that animals *know* anything. All we can be sure of is whether information changed an animal's behaviour or not; we cannot know what goes on in the animal's head. The other meaning of 'information' is statistical and refers to the probability that an animal's behaviour will be changed by a signal. If the behaviour of one animal increases the predictability of the behaviour in another then information has been transmitted. It is this statistical sense of the term 'information' that most researchers now concentrate on.

There is also a problem with the idea that communication can always be measured by its effects on other animals. It is possible that some messages can be received with no outward change in behaviour. For example, if someone tells their dog to 'stay' then the dog does nothing so presumably the command to stay has been communicated. One way of avoiding detection by predators is to remain immobile and so an alarm call may result in no activity despite the fact that information has been sent and received.

To overcome some the problems in defining terms like 'information' and 'signal' Dawkins and Krebs (1978) suggested an alternative view of communication which proposed that the primary function of communication was not to transmit information but to manipulate other animals. Therefore Marion Dawkins (1995) argues that there are two selection pressures acting on signals, one for information transfer and the other for manipulation. It is difficult to arrive at a succinct definition of communication that everyone would agree with, but the following list suggested by Grier and Burke (1992) covers the important elements:

1 Communication requires a signal (coded information), a sender and a receiver.
2 The process benefits the sender but may or may not be advantageous to the receiver.
3 The sender and receiver are usually the same species but they can be different species.
4 Signals can have a number of functions depending on the context and the receiver.
5 Receiving signals *potentially* changes the behaviour of the receiver but we can only be certain that communication has taken place if behaviour changes.

Animals use a variety of means of communicating, including the use of visual, auditory and olfactory signals. It is worth reiterating that some animal communication cannot be detected by humans without using specialised equipment (e.g. some sounds produced by bats or dolphins).

Review the following examples and note whether they are examples of communication:

1 An ant releases a chemical along a trail and other ants follow the trail.
2 The sound of gunfire causes a flock of crows to fly away.
3 A rabbit runs into its warren on hearing the alarm call of a blackbird.
4 The flash of 'eye patterns' on a moth's wings frightens off a predator.

Visual communication

Some of the most interesting and beautiful sights in nature are the displays of animals. These displays consist of conspicuous stereotyped movement and often involve changes of colour and brightness to emphasise them. A vivid example of the use of colour in communication is the display of the peacock's tail that is used to attract peahens. The spectacular courtship display of the great crested grebe is a good example of a stereotyped sequence of movements. Each part of the display was named at the beginning of the twentieth century and include the 'cat display', 'penguin dance' and 'head shaking' (Huxley, 1914). The grebe 'dance' is an example of a ritualised display in which the behaviour of one animal triggers a response in a second animal that in turn triggers a further response in the first, and so on. Another famous example is the 'zig-zag' dance of the male stickleback described by Tinbergen (1951). The male stickleback performs his dance in front of a female and, provided she responds to the dance, leads her to his nest where she lays her eggs. However, if the female does not respond the male stops the dance. The use of visual signals is not restricted to one-to-one communication – some social animals use visual signals as warnings to many other members of the group. For example, rabbits have white rump areas and as one individual runs for safety the white flashes alert other members of the warren to danger, even if they are foraging some distance away.

Visual displays are used as signals for many types of information in a variety of species (including mammals, birds, amphibians and fish). One group of animals that seems to have a complex system of communication is the cephalopod molluscs (squid and octopus). Cephalopods have good vision, well-developed brains and have flexible bodies and colour cells in their skin. They are able to change shape and colour rapidly and dramatically and have been found to use in excess of thirty-five different visual communication patterns (Moynihan and Rodaniche, 1977). According to Moynihan and Rodaniche (1977) it is possible that cephalopods can send more than one message at a time and they claim that one squid can send up to four different messages to four squid by changing different parts of its body.

Not all visual communication involves vivid displays or complex, stereotyped sequences of movement. In some animals, such as the social primates, posture and gesture seem to act as important visual signals. For example, rhesus monkeys use many different gestures and postures to communicate. The estimates of the number of signals used vary according to the definition of communication and range from 22 using a narrow definition (Hinde and Rowell, 1962) to 50 using a broader one (Altmann, 1962). Chimpanzees use a posture to indicate a threat or aggression. This consists of standing upright and raising the fur, and has the effect of making the animal seem larger.

Strengths and limitations

In some conditions visual communication can be very effective, and Grier and Burke (1992) have pointed to a number of advantages of visual communication over other forms. Firstly, visual signals are transmitted very quickly (at the speed of light). Secondly, the displays of an individual can convey a large amount of information. For example, a courtship display may indicate a readiness to mate but it can also reveal information about status, condition, age, etc. Finally, visual information is highly directional, allowing the source to be detected, or hidden, easily.

There are some conditions that restrict the use of visual communication since it can only be used when animals can see each other. If there are obstacles between animals, such as the undergrowth of a forest, then visual communication becomes ineffective. Furthermore, visual communication requires light and cannot be used in dark conditions (at night, in deep oceans, caves, etc.). Some visual signals can be 'expensive' in terms of the energy they require or the risks they cause (e.g. a peacock tail). Finally, Britta Osthaus (personal communication) points out that for the intended recipient of visual communication to receive a signal attention must be directed towards the sender. This is in contrast to auditory or olfactory communication, which do not require attention to be directed at the sender. You can look away or close your eyes, but you cannot switch off your hearing or sense of smell.

Auditory communication

Auditory signals are a potentially rich source of information and are used in communication in a wide range of species. Sounds are very flexible and can be modulated (pitch, loudness and duration) to produce a variety of complex messages. The range of frequencies, measured in hertz (Hz) or cycles per second, used in animal communication is enormous. Humans can detect sounds from about 20Hz to 20,000Hz but some animals, such as bats and dolphins, use ultrasound (which is too high for humans to detect) to communicate. Other animals, such as pigeons, can detect infrasound (which is too low for humans to detect), but little is known about its role in communication.

One group of animals that produces a rich and varied range of sounds is birds. Birdsong seems to have a number of functions in communication, but four main roles can be identified: to act as a territorial warning to rivals, to attract a mate, to identify individuals and to act as an alarm. The songs of many male birds seem to act simultaneously to defend territory and to attract a mate (Catchpole, 1984). The role of birdsong in communicating territoriality was demonstrated in a series of studies of great tits by Krebs (1984). Krebs removed resident pairs of great tits from their territories and then either left the territory vacant or put in loudspeakers playing the song of a great tit. He waited to see how long it took for the territory to be occupied by other great tits. He found that it took longer for the territory with the great tit song being played to be occupied and concluded that the song acts as a 'keep out' signal. Some birds live in large colonies and here the problem can be finding a particular nest site. There is evidence that gannets call while flying above their colony and use the return call of their mates to pinpoint their nest (Thorpe, 1961).

The alarm calls of birds have been widely studied; for example, a series of well-controlled studies have shown that chickens make and respond to different calls for different types of predators (Evans, 1997; Evans and Marler, 1995). Chickens make one type of alarm call for aerial predators and another for ground predators, and they behave in different fashion. On sighting an aerial predator they crouch and look up at the sky, whereas they stand erect and look from side to side when they spot a ground predator. Experiments revealed that if one chicken was shown a video of a hawk they gave the appropriate call and a chicken in another cage showed the behaviour for an aerial predator. If a video of a racoon was shown to the first chicken it gave the call for a ground predator and the behaviour of the second chicken changed.

Some mammals also use auditory communication, and a number have been studied because of the complexity and variety of signals they use. One group that excites a lot of interest is the cetaceans (whales and dolphins) because they use a wide variety of sounds to communicate. In contrast to light, sound can travel great distances in water and this makes it a good mode of communication for marine animals. Whales produce elaborate 'songs' that consist of a number of themes that can be hours long. These songs seem to be signals that are important for social interaction (e.g. Tyack, 1983). Dolphins are generally social animals living in fairly large groups, and they use a wide range of sounds for different purposes. The complexity of the auditory signals used and their apparent intelligence has led to a debate about whether dolphins use a form of language (see pp. 80–81). Another mammal that has been studied because of its range of calls is the vervet monkey. Vervet monkeys use different alarm calls in the presence of different predators: eagles, leopards and snakes (Seyfarth *et al.*, 1980; Seyfarth and Cheney, 1986; see Chapter 5, p. 70, this volume).

Strengths and limitations

There are a number of advantages of using auditory communication. Firstly, it is a very flexible mode of communication that allows for lengthy and complex messages (Grier and Burke, 1992). This flexibility permits sounds to carry a great deal of information and a variety of information (e.g. different calls of the vervet monkeys). Secondly, sound can pass over, around or though many things that block visual information (Bright, 1984). For example, sounds can be heard in the dark or in dense vegetation when visual signals are obscured, and sound can travel through water, wood or earth whereas light cannot. Bright (1984) also notes that sound signals do not linger in the environment and are over quickly.

Grier and Burke (1992) note a number of problems of auditory communication. One major problem is that auditory signals fade quickly and need to be produced repeatedly, and, since they are non-discriminatory, predators can hear them. Another problem with sound signals is that they are prone to distortion and interference (from other sounds, echoes, etc.). Since sounds fade quickly in air they can only be used over relatively short distances (although complex, low-frequency signals can be sent in water over large distances).

Olfactory (chemical) communication

The use of chemical messages, or pheromones, is probably the most universal type of communication, and even animals that rely on other modes of communication also use pheromones. Even very simple organisms such as social amoebas use chemicals to communicate (Bonner, 1969), but most research has concentrated on insects and mammals. Chemicals differ in their properties and can be used for different purposes. Highly volatile chemicals are useful for alarm signals because the signal disperses quickly. It would be of little use to have a long lasting chemical to signal danger because recipients would not know whether danger was present or past. On the other hand, long-lasting chemicals are very useful to mark territories since they continue to act as a signal after the signaller has gone.

Insects tend to use volatile chemicals, and most social insects (such as ants and bees) have many glands that release different pheromones (Wilson, 1965). Ants live in nests that may contain many thousands of individuals and they use pheromones to control their social behaviour. Ants use pheromones to recruit other individuals to exploit a food source, to recruit individuals for defence of the nest, to warn of danger, to identify what nest individuals are from, and to attract mates. When danger threatens, some species of ant release different chemicals to elicit different responses (Ridley, 1995). Studies of pheromones have shown that insects can be very sensitive to one specific chemical. Female silkworm moths release a pheromone called bombykol and experiments have shown that males are sensitive to a few molecules and can detect females several kilometres away (Schneider, 1974). However, the male responses are specific to bombykol only and other molecules do not stimulate them, even if they are very similar in composition and structure.

Mammals also use pheromones, but they typically use less volatile, more long-lasting chemicals to mark territory or to attract mates. For example, in a study of six wolf packs Peters and Mech (1975) found that the territory boundaries were clearly delineated by urine scent marks. The territories are large, and as it can be several days before the wolves renew the marks at each point the chemical has to be long lasting. Some mammals have a number of glands that produce and release a variety of pheromones. Black-tailed deer, for example, have six different glands that produce pheromones (Muller-Schwarze, 1971).

Strengths and limitations

Grier and Burke (1992) suggest that chemical communication has a number of advantages over other forms. Firstly, chemicals can be used to send signals even in the absence of the signaller (e.g. territory marking). Secondly, in the air chemicals have a great range and can transmit around obstacles. Finally, pheromones often need specific receptors to receive them and these are frequently species-specific. The chemical signals are therefore 'private' and remain undetected by predators. For example, the male moths can detect a few molecules of the pheromone released by females but predators such as bats do not detect the chemical.

There are a number of disadvantages of using chemical communication. The use of chemicals to send signals is inflexible since each pheromone conveys only one message. Consequently an animal needs a different gland for each chemical message (whereas a range of auditory messages can be conveyed using one mechanism such as vocal cords). Another problem is that, compared to light and sound, chemicals are a slow mode of communication (Manning and Dawkins, 1992).

Progress exercise

Dogs have good senses of hearing, vision and smell. Write a brief description of how you might test the role of each of these senses in communication between dogs.

Other factors in communication

The description of animal communication so far has assumed that animals communicate by using one type of signal at a time (i.e. using one mode or channel) and that the information that is being sent is honest. Neither of these assumptions is always correct since animals often use several types of signal to communicate simultaneously (**multimodal communication**) and there has been some debate about the honesty of signals. Another important aspect to mention is the context in which the communication takes place.

Multimodal communication

The description of visual, auditory and olfactory communication above gives the impression of discrete communication signals that use one channel or mode to transfer information. However, a lot of animal communication is multimodal as it uses more than one channel or mode simultaneously. For example, the visual threat display of the chimpanzee described on p. 52 is accompanied by a series of hoots which become louder and more rapid before they end in a screech. The combination of visual and auditory information is common in mammals and birds and can also include other channels such as touch. One problem of studying multimodal communication is that it is difficult to analyse which signal changes the behaviour of the recipient, or whether it is the combination of signals that is important.

Context

One assumption that has been implicit in the discussion of communication so far is that it can be studied in a controlled isolated fashion. However, communication, by its very nature, takes place in a social setting, and the social environment or context in which communication takes place can influence responses. Some cognitive ethologists argue that a lot of animal communication cannot be understood without reference to context since the context provides further sources of information. For example, Smith (1996) in a discussion of the role of communication and context points out that: 'It is the evaluation of information in context that is centrally important to studies of both cognition and communication. Communication cannot be understood without taking into account the mental operations that integrate information from many sources' (p. 243). Smith refers to this integration of information from the signal and the environment as context-dependent responding, and describes how context both enriches communication and makes it more flexible. Some communication is not context-dependent, as some responses seem to be triggered by a single stimulus ('releasers'), but there does seem to be context-dependency in more complex social interactions. It may be that the context provides some expectation that certain events, and subsequent responses, are more likely than others (Smith, 1996). These expectations are a form of mental set and would probably be based on experience of previous similar situations.

Deception

There has been a great deal of debate about the 'honesty' of signals in animal communication. The word 'honest' is used to refer to whether the signal accurately reflects reality; it does not necessarily imply a conscious decision to deceive another as it would in human communication. When communication between animals has mutual benefits then there is no reason to suppose that the signals sent are dishonest. However, when there is a conflict of interest (for example between two rival males) there seems to be potential for dishonesty. Threat displays rarely involve much physical contact, so animals could 'bluff' and gain a food source or a mate. Thus the use of dishonest signals would seem to benefit the individual, yet most studies tend to show that dishonesty is rare. Ridley (1995) suggests there are at least two reasons why dishonest signals have not evolved. One is that some signals are directly related to fighting strength and cannot be faked. Red stags, for example, roar to deter rivals, but roaring is tiring and can only be sustained at a high rate by genuinely fit and healthy stags (Clutton-Brock and Albon, 1979). Any stag that heard another which could roar faster and longer would have an indication that its rival was stronger. The second reason is related to the 'cost' of the signals. If the signal has a high cost to produce (perhaps in energy or biological resources) then a strong, fit individual would gain more than a weaker, unhealthy one. This is because it would use relatively more of the resources of the weaker individual. Stronger individuals can therefore 'afford' more powerful signals.

There is some evidence that signals can be used in a dishonest way, particularly amongst primates. For example, Byrne and Whiten (1988) describe a number of instances of primates using signals in an apparently dishonest way to gain advantage. However, these examples raise the issue of intent and could be used as evidence of prevarication in animal communication (discussed in the next chapter: see pp. 64–65).

Summary

The concept of animal communication is difficult to define and this has led to some debate and confusion about its exact nature. Most definitions stress that animal communication is the transmission of information between animals that uses signals which have evolved for that purpose. Many also include the notion that communication is usually advantageous to the sender. Animals use a variety of modes of communication, including visual, auditory and olfactory. Visual communication can involve conspicuous displays or more subtle forms of posture or gesture. It is quick, directional and can convey a lot of information, though it is unsuitable in dark conditions or where vision can be obscured (in undergrowth for example). Auditory signals can be a flexible form of communication that can be used in conditions that restrict visual communication (darkness, in water, etc.). However, in air auditory signals fade and distort rapidly and the signals are short lived. Predators can readily detect auditory signals. Olfactory communication is widely used and has the advantage of sending signals after the signaller has left and can remain undetected by predators. However, olfactory signals tend to be inflexible and slow. A lot of animal communication is multimodal since it uses a number of signals simultaneously. Some signals convey a different message, depending upon the context in which they are sent, and a full understanding of communication requires a study both of signals and of situation. Most signals seem to be honest, and this seems to be explained by evolutionary selection.

Review exercise

Complete the following table:

Mode of communication	Key advantages	Key problems
Visual		
Auditory		
Olfactory		

Further reading

Grier, J.W. and Burke, T. (1992) *Biology of Animal Behaviour*, Dubuque, Ia.: Wm. C. Brown Communications. This book has a very thorough chapter on animal communication that deals with definitions of communication and various types of communication.

Ridley, M. (1995) *Animal Behaviour: A Concise Introduction* (2nd edn), Oxford: Blackwell. The chapter on communication deals with the topic in a clear and interesting fashion.

Animal language

Introduction

For the past few decades there has been a debate about whether animals have, or can acquire, the ability to use language. At times the argument between those who are convinced that animals can use language to a degree and those who believe that animals show no signs of using language at all became quite heated. Lewin (1991) characterises these two positions as the *continuity school* and the *discontinuity school*. The continuity school views the ability to use language ability as a continuum, with human language at one end and a complete lack of language ability at the other. This school of thought claims that, although no non-human animal has the same level of language ability as humans, some animals demonstrate most of the major features of a language but do so to a lesser degree than humans do. The discontinuity school, on the other hand, views the use of language as uniquely human and as an ability that sets us apart from other animals. They regard any 'evidence' of language in animals as an illusion and believe that all that the evidence shows is that animals can learn complex sequences of action to gain rewards.

There are a number of reasons why the debate causes such disagreement, but one of the principal factors is the sophistication of human language. The use of language is one of the most complex human skills, and agreeing an exact definition of language is problematic because it has many features and functions. In the absence of a clear and precise definition of human language it is difficult to assess the language skills, if any, of animals. Also, as with most questions about animal cognition, there is a problem of methodology and interpretation. We cannot expect animals to use language in exactly the same way as humans, but what can we interpret as the equivalent of human language? Savage-Rumbaugh and Brakke (1996) point out that because animals do not speak they must be taught a communication system (such as signing or the use of symbols) and that this introduces methodological problems that make comparisons between species more difficult. They claim this difficulty shifts the focus away from the 'functional aspects of symbolic representation' to the mechanisms of the communication (i.e. the emphasis shifts to *how* communication takes place rather than on *what* is communicated).

The question of whether animals use, or can acquire, language centres on what features of language are regarded as the essential characteristics. Thus the starting point of this chapter is a consideration of the characteristics of language.

Characteristics of language

Human language is a complex form of communication that defies simple definitions. It can take many forms and, in addition to the thousands of spoken and written languages, there are various types of sign language and numerous ways of signalling language (Morse code etc.). Language can be used to communicate ideas, emotions, to lie and to discuss the impossible. One way to encompass the complexity of language is to identify a range of characteristics that describe its essential features. Hockett (1959, 1960) took this approach and described 16 'design characteristics' of human language (see Box 5.1).

Box 5.1 Hockett's design characteristics of language

1 *Vocal/auditory character*: sounds are used to transmit language from one person to another.
2 *Broadcast transmission and directional reception*: language is broadcast and can be heard by anyone in range. Also the receiver can tell where it is from.
3 *Rapid failing*: the signal fades rapidly.
4 *Feedback*: the sender hears the language they produce as they speak.
5 *Interchangeability*: language is a two-way process that involves both the sending and receiving of signals.
6 *Specialisation*: language is not a by-product of some other biological function, it has a special function for communication only.
7 *Semanticity*: language conveys meaning and refers to features in the real world.
8 *Arbitrariness*: the signals do not resemble what they represent.
9 *Traditional transmission*: language can be transferred from one generation to the next.
10 *Learnability*: language can be learned.
11 *Discreteness*: language is organised into discrete units (e.g. words).
12 *Displacement*: language can used to refer to things that are not present in space (here) or time (now). These are known as 'spatial displacement' and 'temporal displacement' respectively.
13 *Duality of patterning*: there are two patterns in speech; sounds are arranged into words and words are arranged into sentences.
14 *Productivity*: language can be used to produce an infinite variety of new messages.
15 *Prevarication*: language can be used to talk about the impossible or to deceive. Language can also be used to withhold information.
16 *Reflexiveness*: language can be used to talk about (or reflect on) language.

Some of the design characteristics (1 to 4) are only applicable to human spoken language and are not relevant to all forms of language (including human sign language). However, some of the design characteristics are seen as essential features of any form of language and are useful criteria to assess claims about animal language. Brown (1973), for example, regards **productivity**, **displacement** and **semanticity** as the three most important characteristics of language. Some of these characteristics are significant because they seem to distinguish human language from simple animal communication. For example, human language is productive because we can use **grammar** to combine and recombine words in an infinite variety of sentences. Furthermore, because we follow grammatical rules other users of the language can recognise the meaning of each sentence immediately – even though they have never heard the same sequence of words before. On the other hand, most forms of animal communication are very restrictive: a few sounds, gestures or odours are used to signal a fixed message. Animal communication has also been described as reflexive or stimulus-bound; in other words it is used to send signals about what is physically present. Alarm calls, for example, are made only in the presence of a predator and not at any other time. However, human language is capable of displacement; we are able to discuss objects and events that are not present either in space or time. You may discuss the danger posed by great white sharks even though the nearest one may be several hundred kilometres away. Alternatively I could inform you about the danger I was in yesterday or even the danger I *may* be in tomorrow. Human language is not bound by the stimulus; it is voluntary and could be described as 'intentional communication'. Semanticity is another important characteristic because the primary function of human language is to communicate some meaning such as ideas, information or feelings between people. It is not possible to study ideas or feelings in animals, but Shettleworth (1998) describes how it is possible to study semanticity by looking for evidence or signals that are **functionally referential**. She defines these as signals that are made in response to a specific object or event and that elicit a consistent behaviour even in the absence of the object or event. For example, if every time I yelled 'fire' you were to run to the exit, even though you had not seen any fire, the sound 'fire' would be acting as a functionally referential signal. Typically, in animal communication this is not the case since signals and responses are general and stimulus-bound. For example, an alarm call is made in the presence of any danger and usually triggers a general, not specific, response.

Another feature of human language that seems different from animal communication is **prevarication**, which is the ability to talk about absurdities, the impossible, or even to use language to deliberately deceive others (to lie). You could tell a friend that there was a leopard behind them but this would be both a lie and absurd (in the vast majority of cases!). However, since most animal communication seems to be in response to a stimulus (i.e. it is reflexive) it seems to be incapable of prevarication.

There are two ways of examining whether animals are capable of using language or not: firstly, by studying natural animal communication to see whether any of them fulfil the criteria for language; secondly, by trying to artificially teach animals a form of language to assess their language abilities.

Progress exercise

Compare human language and animal communication by completing the following table:

Design characteristic	Human language	Animal communication
Semanticity		
Productivity	Can use grammar to produce an infinite variety of messages	
Displacement		Is reflexive and is made in response to what is present
Prevarication		

Natural animal language

Honey-bees

Honey-bees have a remarkable system of communication that is used to indicate the locations of food sources. It had been recognised for some time that when one bee found a food source a large number of bees would soon begin to exploit it. However, the mechanism by which bees recruited fellow workers was not known until a series of studies by von Frisch (1950, 1974) revealed that they used dances to transmit information. What he discovered, in an animal with a small nervous system, seems to be a complex form of communication (see Chapter 7, pp. 110–113). When a worker bee returns to the hive after discovering some food she first gives some of the food to other bees and then performs a dance on the vertical surface of the honeycomb. If the food source is near (within approximately 50 metres) the bee performs a **round dance**. This consists staying more or less in the same spot and turning left and right alternately for about 30 seconds. This seems to stimulate other bees to find the local food source.

If the food source is further away the bee performs a more elaborate dance called the **waggle dance**. During this dance the bee performs a figure of eight movement by moving in a straight line, circling to the left to the start of the line and then following the line again before turning right. As the bee runs along the straight-line part of the dance it waggles its abdomen. Von Frisch discovered this by using a technique to study the behaviour of bees after they had found a food source. He placed dishes of sugar at various locations around the hive and waited for a bee to discover it. When a bee arrived von Frisch marked it with paint and then observed the bee's behaviour back at the hive. He found that the waggle dance gave an indication of both the distance and direction of the food source. The distance of the source is linked to the length of the straight run of the dance – the further the distance the longer the straight line (although, since this determines the number of waggles and the time taken to perform each figure eight, the exact signal is not clear). Von Frisch found that the angle of the straight part of the dance was always at an angle relative to vertical that was used as an indication of the direction of the food. The angle corresponded to the angle of the food source from a straight line between the hive to the sun (Figure 5.1). Thus if the food source was in a direct line between the hive and the sun the straight line of the dance was vertical. If the food source was 15 degrees to the left of a line between the hive and the sun the straight line was 15 degrees to the left of vertical.

EVALUATION

The waggle dance shows many of the characteristics of language such as displacement (the dance indicates something that is not present), interchangeability (bees can both perform and respond to the dance) and

(a)

(b)

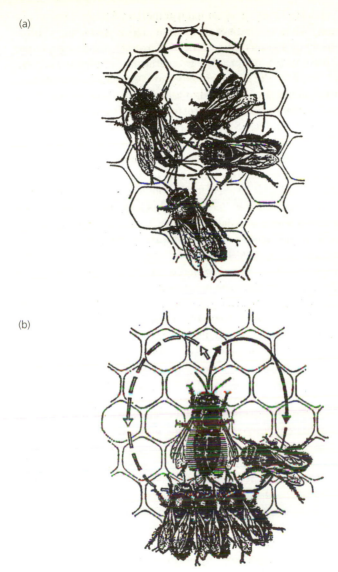

Figure 5.1 **The dances of honey-bees**

    (a) The round dance
    (b) The waggle dance

Source: Based on von Frisch (1974); copyright © The Nobel Foundation, 1974.

specialisation. However, there are many important features of language which are not shown in the bee dance. The dances show no evidence of productivity, traditional transmission or learning. Bees cannot use the dance to produce different messages apart from distance or direction. Bees have dance 'dialects' which vary amongst species (von Frisch, 1962), and if larvae are displaced to a hive of a different species as adult bees they behave as they would in their natural hive not the adopted one (Gould and Gould, 1988). This suggests that there is no learning or traditional transmission of the communication. There is also a question about the function of the waggle dance in communication. An alternative explanation of how bees find food sources is that they use odours, both from the food itself and other distinctive odours near the food source, to navigate (Wenner and Wells, 1990). This hypothesis suggests that the dance merely acts as a signal to other bees to take note of the odours. Although he acknowledged the role of scent in finding nearby food sources following a round dance, von Frisch (1974) has dismissed the possibility that scent was used to find distant sources after a waggle dance. He pointed out that if hives are turned on their side so that the honeycomb is horizontal the waggle dances become confused and few bees find a scented food source. If the hive is returned to its normal position many new recruits find the source, suggesting that it was the dance not the scent which indicated direction.

Cetaceans

There has been great interest in whether cetaceans use a language in their natural habitat since, like humans, they use complex auditory signals, are social and they have large brains. Many species of whale produce intricate sequences of sounds that have been called 'songs'. The humpback whale, for example, produces a song that has an elaborate structure. The basic elements of the songs are units that are combined into phrases, a number of phrases being linked to make a theme. A single song consists of up to ten themes, and analysis of the songs shows that there are differences between humpback whales from different areas. The songs are used in courtship and seem to have a social function. Tyack (1983) found that when a recording of a song was broadcast via speakers on a ship other whales stopped singing and approached the ship. They seemed to be searching for the source of the songs.

Dolphins use a wide variety of sounds which can be divided into two groups: pulsed and unpulsed sounds. The pulsed sounds are high-frequency clicks and bursts and the unpulsed sounds are whistles and squeaks. This variety of sounds (which include ultrasound) adds to the problem of studying the possibility of a natural dolphin 'language' because it is difficult to isolate which sound is linked to which behaviour (if any). However, there is evidence that each dolphin has its sequence of clicks that it uses to identify itself. In addition, schools of dolphins are very vocal, suggesting that the sounds have social functions. An observation of two captive dolphins suggests that the sounds could possibly be used in a sophisticated way (Bastion, 1967). The dolphins were placed in separate tanks that contained identical apparatus, and although they could not see each other they could hear each other. One dolphin was then taught to push a paddle to gain a reward, but the other was not. However, soon after the first dolphin had learned to push the paddle the second began doing so as well. During the training period both dolphins produced a lot of sounds. This may be a coincidence, but it raises the possibility that some information about a novel situation was communicated. If so this would be evidence of animal communication with language-like properties.

EVALUATION

There are difficulties in studying cetaceans because their environment is so alien to human investigators. Even if the range of sounds produced by dolphins and whales are a complex form of communication it would be difficult for humans to interpret or to understand. This is because, although it is possible to record the sounds in water over large distances, it is not easy to see the effects on a recipient's behaviour (or indeed who is the recipient). Pearce (1997) notes that little is known about the function of these signals and that there remains a possibility that, in dolphins, they may be a system of communication that approximates language. However, he argues that the complex communication involved in language should be accompanied by correspondingly complex social interactions and that there is little evidence of this from naturalistic evidence.

Primates

Monkeys and apes use a range of sounds and gestures to communicate; like cetaceans, they have large brains, are social and are able to learn readily. It is therefore possible that they may use a communication system akin to language. One type of monkey that seems to have a complicated system of communication about predators is the vervet monkey. As mentioned in the previous chapter vervet monkeys have a number of different calls for different predators (eagles, leopards and snakes). Furthermore, the different calls produce different reactions which are appropriate for the type of predator they represent. On hearing the eagle call vervet monkeys retreat from the tops of trees and look up (since eagles strike from above). The response to the leopard call is to run to trees and climb (leopards cannot reach the thin branches that will support monkeys). The snake call causes the monkeys to stand on their hind legs and look at the ground. When the different calls are played over loudspeakers they trigger the appropriate responses even in the absence of a real predator (Seyfarth et al., 1980; Seyfarth and Cheney, 1986). This seems to be good evidence of a functionally referential signal (see p. 64) and indicates semanticity.

It appears that some aspects of the vervet calls are learned. Infant vervets make alarm calls in a roughly appropriate manner, but only learn to discriminate between different objects during their first four years (Seyfarth and Cheney, 1986). For example, the infants make eagle calls at the sight of many types of bird, as juveniles they only make the calls at the sight of birds of prey, and then as adults they only make the calls when they see the eagles that prey on them. There is a similar process in gradually discriminating between mammals in general and leopards. Since this discrimination seems to be learned from the other members of the troop this is evidence of learnability and traditional transmission.

Byrne and Whiten (1987, 1988) describe a number of instances of baboons using signals in an apparently dishonest way to gain advantage (or evidence of prevarication). For example, a juvenile baboon watched as an adult dug up a large root and then, after looking around, screamed loudly. The juvenile's mother, who was dominant in the troop, then ran over and chased the other adult away. The juvenile then picked up the root and ate it. The juvenile's scream appears to be a signal designed to mislead the mother and provoke an attack. However, this and other examples described by Byrne and Whiten raise the question of interpretation of intent of animals. Are they deliberately using communication to deceive or are there simpler explanations? For example, an alternative explanation of the juvenile baboon's behaviour is that it is a conditioned response that occurred because of past experience. If the juvenile had approached an adult who was eating and had then been threatened a normal response would be to scream. The scream could trigger an attack by the mother and the juvenile would be left with the food; it would be reinforced for screaming. Byrne and Whiten (1987) discuss this possibility, but note that the juvenile showed the behaviour repeatedly only when its mother was not looking. They believe this could indicate an 'intent' to deceive.

EVALUATION

Studies of communication between monkeys in their natural habitat indicate that they demonstrate a number of the features of human language. The auditory signals of the vervet monkeys seem to show semanticity, learnability and traditional transmission. There is also evidence of prevarication in baboons, but neither species (nor any other monkey or ape) shows evidence of productivity in their natural habitat. The problem with such studies in the natural environment is the lack of control (e.g. there is no record of the past behaviour), and therefore any conclusions rely on interpretation.

Conclusion

Some natural animal communication involves complex transfer of information which can show *some* of the characteristics of human language. For example, bees use a dance to indicate a food source and show evidence of displacement. Vervet monkeys use different sounds to represent different predators and show semanticity and arbitrariness. However, any natural animal communication only shows a few features of human language; none show the *full range* of features that distinguish language from communication. For example, neither bees nor vervet monkeys show any evidence of grammar, which is vital for generating novel messages (productivity). No system of natural animal communication seems capable of reflexiveness (using language to discuss language), or of any discussion of a similarly abstract nature. Another factor that has to be considered when studying natural animal communication is the problem of interpretation (e.g. are instances of 'deception' in natural communication examples of prevarication or conditioning?).

Using Hockett's design characteristics of language (p. 63):

1 Use examples to illustrate which key characteristics are shown in natural animal communication.
2 Which key characteristics are not shown in any of the examples of natural animal communication?

Can animals learn language?

A second approach to investigate the question of whether animals are capable of language is to use a more controlled setting and to try to teach animals to use or understand a language. However, to ask 'Can animals learn language?' has been described as the wrong question (Shettleworth, 1998), the reason being that no species seems to have exactly the same abilities as an adult human. Shettleworth argues that a better question is 'What aspects, if any, of human language can be acquired by another species?', and that the key characteristic to investigate is productivity. In other words, the focus should be to find whether there is any evidence that animals can learn syntax or grammatical rules in order to combine signals to produce new meanings.

Apes

There have been a number of attempts to teach apes language using a variety of techniques. Two early studies tried to teach chimpanzees to speak in the same way a human child would by raising them in a home and exposing them to language. One chimpanzee, Gua, was raised with a human child by Kellogg and Kellogg (1933) but never learned to say a single word (although there was evidence that she did understand some words). A similar study by Hayes and Hayes (1951) with a chimpanzee called Vicki was also a failure – even after intensive training Vicki only learned to say four words. Both these attempts failed because they concentrated on trying to teach a spoken language that the chimpanzees were not equipped to produce.

A breakthrough in the research into animal language seemed to occur when Gardner and Gardner (1969) used a different technique with another chimpanzee, Washoe. Although chimpanzees are ill-equipped to produce any spoken language they are dextrous, and Gardner and Gardner argued that a more realistic goal would be to teach Washoe American Sign Language (ASL). They created a signing environment in which nobody spoke but all the humans signed to each other and to Washoe. They taught Washoe to use signs by moulding her hands to the correct position and reinforcing correct signs (e.g. operant conditioning). Washoe learned signs slowly at first but after four years' training she could use about 160 signs that included nouns, verbs and pronouns. She was able to use combinations of signs, starting with two signs and then progressing to four- or five-sign 'sentences'. One famous example is the novel sequence 'baby in my drink' which was produced in response to being shown her cup with a doll in it. Washoe also seemed able to respond to questions posed by her trainer and to produce novel combinations of signs. For example, when shown a swan and asked 'what's that?', Washoe signed 'water bird'. Some years after the start of the study (and after the death of her own baby) Washoe adopted a ten-month-old infant, Loulis. The trainers did not use signs in front of Loulis yet he learned to use a number of signs. Some of these were learned by imitating Washoe, but there was also evidence that she moulded his hands into the correct position (Fouts *et al.*, 1982). This seems to be evidence of traditional transmission (see p. 63).

A similar study using ASL was carried out using a gorilla, Koko (Patterson, 1978). One major difference is that Patterson used both signs and spoken language in the gorilla's presence. Koko learned to use many signs, and she responded to both signs and spoken words. She appeared to use the signs creatively to refer to new objects; for example, she signed 'bottle match' to refer to a cigarette lighter, 'white tiger' to refer to a zebra and 'elephant baby' when shown a Pinocchio doll (Patterson, 1980). Koko was later joined by another gorilla, Michael, and showed displacement by referring to him when he was not present. She became something of a media celebrity following the death of her kitten (which she apparently had named 'All Ball'), because of her apparent distress. No matter how these signs about her distress at the loss of her kitten are interpreted, they are indications of displacement since she was referring to something that was not present. (More information about Koko can be found on a website: www.koko.org)

The excitement generated by these early studies was tempered by the publication of *Nim* by Terrace in 1979. (The name Nim was short for Neam Chimpsky, which is a pun on the name of the linguist Noam Chomsky who believed that language is uniquely human.) Nim was another chimpanzee who was systematically taught sign language; during the study he learned 125 signs, and a record of over 19,000 'utterances' he made was kept. Initially these utterances seemed to be much like a child's language development as single signs were combined into two sign sentences with a consistent 'grammar'. However, the similarity ended there and Terrace and his colleagues found little evidence of Nim showing any ability to use language. Firstly, Nim remained stuck at the two-sign stage and any increase in the length of signing was the result of repetition (e.g. 'banana Nim banana Nim'). Secondly, when using more than two signs Nim showed no evidence of use of grammar and did not produce them in a consistent manner. Thirdly, unlike a human child, Nim did not take note of other signs and respond (or 'converse') but would make signs that interrupted others. Finally, detailed analysis of the video records of Nim signing revealed a tendency to repeat the signs made by the trainers. Terrace concluded that Nim did not show evidence of language learning but that he either simply imitated signs or used simple sequences for reward.

Use of symbols

One of the drawbacks of using sign language with primates is the problem of control. It is difficult to analyse whether a sign is spontaneous or is an imitation of signs from the trainer, and any sign produced by the animal is open to interpretation. Several alternative approaches to teaching animals language have been used that allow more control and objectivity. One approach was adopted by Premack (1971,1976), who used plastic shapes, which varied in shape and colour, to represent words with a chimpanzee, Sarah. This has the advantage over signs of being easy to interpret and quantify. Sarah learned to use about 130 of these shapes to represent nouns, verbs, adjectives, etc. and she was required to put them in correct order on a board to make sentences. She seemed to be able to arrange the shapes in the correct order and was able to follow instructions which the trainer placed on the board. For example, if the trainer gave the instruction 'Sarah give me apple', Sarah would do so. She was also able to answer questions about the relationship between objects correctly. If she was shown a green card on top of a red card and asked if the green card was on top she would answer 'no'.

Rumbaugh (1977) adopted a slightly different approach and used symbols (termed 'lexigrams') on a keyboard which was connected to a computer; the animals were required to use the symbols in the correct sequence (to form a language called 'Yerkish'). This technique has the added advantage of recording all the sequences that the animal might make. The first to be trained in Yerkish was a chimpanzee called Lana who used the symbols primarily to ask for things (e.g. 'Please machine give apple'). However, many of these sequences can be explained in terms of conditioned responses (to obtain food etc.). Later studies using Yerkish with two other chimpanzees, Sherman and Austin, emphasised the social use of language and understanding of meaning (Savage-Rumbaugh, 1986). They have been trained to ask for tools to open food containers using lexigrams and to request and share food with one another. They also showed evidence of displacement in tests where they were shown a collection of food and drink items in one room and required to use the lexigram in another room to state which one they would choose from the display. When they returned to the first room they were permitted to eat or drink the item if they picked the one they had stated with the lexigram. They were able to do this with 90 per cent accuracy (Rumbaugh and Savage-Rumbaugh, 1994). However, the studies were still open to the criticism that the chimpanzees were simply learning routines to obtain rewards rather than using the symbols as a form of language.

Informal language learning

One of the most interesting studies of language learning in primates is that of a bonobo (or pygmy chimpanzee) called Kanzi (Rumbaugh and Savage-Rumbaugh, 1994; Savage-Rumbaugh and Brakke, 1996). Kanzi is a good example of how serendipity, or happy chance, can help a topic to progress since he was not the original subject of the study. Rumbaugh and Savage-Rumbaugh (see Chapter 7, pp. 119–122) began by studying Kanzi's foster mother Matata with lexigrams, but she failed to learn Yerkish well. However, even though no attempt was made to teach Kanzi, he started to use the lexigram to request and name things; furthermore he showed signs of understanding some spoken English. It was realised that Kanzi had acquired comprehension skills through exposure to language, much as a human child does. He was therefore treated like a young child and was allowed to roam a large wooded area and given stimulating experiences accompanied by humans who both spoke and communicated on a portable lexigram keyboard. After a number of years a series of different carefully controlled tests showed that Kanzi could both understand and produce sentences. For example, he was able to follow simple instructions such as 'Make the doggie bite the snake' or 'Get the apple from the fridge'. The tests used completely novel sentences and precautions were taken to prevent any cueing by his trainers. Kanzi was able to collect items from the correct location even if an identical item was in front of him. So if he were asked to get a pine cone from the fridge he did so even if there was another pine cone in full view. An important difference between Kanzi and the primates in previous studies is that much of his communication is spontaneous and is not made in response to his trainers. A comparative study of Kanzi and a young child suggests that his comprehension skills are equivalent to that of a 2½-year-old child, but his skills in producing sentences are more like those of a 1½ year old (Savage-Rumbaugh *et al.*, 1993). Savage-Rumbaugh and Brakke (1996) suggest that Kanzi uses a simple version of grammar (a 'protogrammar'), and although he does not use language in the same way as an adult human he shows clear signs of language abilities. He shows the key features identified by Brown (1973) of semanticity, displacement and productivity; furthermore he produces sentences with spontaneity. Savage-Rumbaugh and her colleagues have raised other chimpanzees and bonobos in a similar fashion to Kanzi with similar success and, given the failure with older individuals, have suggested that these results may indicate that there is a sensitive period early in primate development when apes need to be exposed to language.

EVALUATION

The various studies of apes using sign language, plastic shapes or lexigrams have shown that they are able to manipulate signals and show

Figure 5.2 **Picture of Kanzi listening to instructions on headphones**
Any response to these instructions is to the language alone since no other
cues are available.

Source: Figure from 'Language in comparative perspective', by D.M. Rumbaugh
and E.S. Savage-Rumbaugh, in *Animal Learning and Cognition*, edited by N.J.
Mackintosh, copyright © 1994 by Academic Press, reproduced by permission of
the publisher.

some of the characteristics of language. All the studies of apes have
shown that they are able to learn signs/symbols to represent objects.
For example, using a double-blind technique Gardner and Gardner
(1969) were able to show that Washoe was able to name objects
correctly. Furthermore the animals seem to be able to show the appropriate responses to signs/symbols and, in some cases, spoken English.
For instance, Kanzi can follow instructions given in English even if the
instruction is novel. Many of the studies have shown that primates are
capable of displacement and can produce and respond to signs/symbols
that refer to objects that are not present. There is evidence of both spatial
and temporal displacement in primate communication. For example,
both Washoe and Koko showed spatial displacement. The day after her
infant was taken away for emergency treatment the first sign that
Washoe made to her trainer was 'Baby'; Koko signed both about her
companion Michael and her kitten when they were not present. Koko

also seems to show temporal displacement in a sequence between her and (Penny) Patterson following an incident when Koko had bitten Patterson:

PATTERSON:	What did you do to Penny?
KOKO:	Bite.
PATTERSON:	You admit it?
KOKO:	Sorry bite scratch. Wrong bite.
PATTERSON:	Why bite?
KOKO:	Because mad.
PATTERSON:	Why mad?
KOKO:	Don't know.

There is also evidence that the primates are capable of prevarication. When Patterson asked Koko about a broken catch to a window Koko indicated that another of her trainers had done it. Finally there is also evidence that primates are capable of productivity, or the generation and understanding of novel sentences using a form of grammar. Lana's responses, for instance, showed she could recognise the difference between 'Lana groom Tim' and 'Tim groom Lana' . Sarah produced and responded to the correct sequence of symbols. Koko seemed to have her own form of abuse for anyone who displeased her – she called them a 'Big dirty toilet'!

The conclusion that apes can learn the to use language has been challenged by Terrace *et al*. (1979) (see Chapter 7, pp. 116–119). They asked 'Can an ape create a sentence?' and, based on their study of Nim and other studies using sign language and symbols, concluded that the answer was 'no' for a number of reasons. After using single words at first, then two-word sentences, the length of sentences that children use gradually increases, and they also use a consistent form of grammar; this did not happen in the studies of apes. Nim, for example, produced an average sentence length of only 1.5 words and one of his longest 'sentences' was 'give orange me give eat orange me eat orange give me eat orange give me you'. This showed a lack of grammar and a tendency to repeat signs, which Terrace *et al*. found to be the norm. Children also engage in language without being prompted and from a very early age they are able to take turns in communication. In contrast the chimpanzees rarely used signs spontaneously and tended to respond to prompts from trainers. Terrace *et al*. also found that the chimpanzees had a tendency to imitate their trainers so that what at first sight seemed to be a sentence was merely a copying of the trainer or was cued by the trainer. For example, when the famous sequence of Washoe signing 'baby in my drink' was carefully analysed the sequence seems to be in response to cues from the trainer. Another common criticism of the primate language studies is that the animals produce the signs because they had been conditioned to do so and that it did not reflect any ability to use a language. In other words they were performing sequences of signs or symbols to obtain reinforcement but the sequence had no meaning for them. For example, if I want to check the weather forecast on the teletext pages I have to press five buttons in the correct sequence. Should this be regarded as a language or a sequence of behaviours to get want I want? However, it may be a mistake to dismiss the use of signals as 'just conditioning'. Shettleworth (1998) points out that: 'Much of the controversy in this area boils down to disagreement over whether the subjects "really" have one or another of linguistic compe- tence – syntax, reference, etc. – or whether their behaviour is "merely" instrumental responding' (p. 562). She claims that it is paradoxical that, while it is recognised that conditioning procedures can produce elaborate representations of the world, if communication behaviour is interpreted as a result of conditioning it is viewed as simple and uninteresting.

Many of the questions raised by Terrace *et al.* (1979) about the ability of apes to use sign language are not relevant to Kanzi and the other bonobos studied by Savage-Rumbaugh. Kanzi does seem to exhibit many of the features that are viewed as essential characteristics of language. He shows semanticity, displacement and productivity, and is able to both comprehend and produce language. Furthermore Kanzi learned language by observation rather than by conditioning and it is difficult to see how his use of symbols can be explained as a simple conditioned response. A key difference between Kanzi and the other primates is that Kanzi was exposed to language at an early age in a social context. Also the emphasis of the research was initially on the comprehension rather than the production of language. The ongoing studies of Kanzi and other bonobos may shed more light on the linguistic abilities of apes.

Dolphins

Most of the work on language learning in apes up to Project Kanzi had tried to assess both the comprehension and production of words and sentences. In contrast the work on teaching dolphins language has concentrated mainly on comprehension. Herman and Morrel-Samuels (1996) suggest there are a number of reasons for this, including the problems of obtaining objective, replicable data from studies that emphasise productive skills, and that there is evidence that receptive skills are likely to be a more valid indicator of language potential. Herman and his colleagues trained two bottlenosed dolphins to respond to sequences of signals (Herman *et al*., 1984; Herman *et al*., 1993). One dolphin, Phoenix, was taught to respond to acoustic signals (short computer-generated noises), while the other, Akeakamai, was taught to respond to gestures (hand and arm movements of a trainer by the pool). Both sets of signals were designed to act as a type of language with lexical components (words) to represent objects, actions and modifiers and a set of rules or syntax to combine the signals (grammar). The words and syntax taught to the dolphins were carefully controlled so that their comprehension of novel sentences could be tested. The words that the dolphins learned allowed a large range of sentences to be generated and the animal's comprehension was then tested by analysing their responses. For example, a sentence that means 'take the ball to the frisbee' should lead to a different response to 'take the frisbee to the ball', even though the same signals are used. They were initially taught two-word signals but learned to respond to novel four- and five-word signals which convey complex instructions. For example, Akeakamai correctly responded to 'ball right frisbee fetch' (which, using the syntax Herman *et al*. created, means take the frisbee on your right to the ball). Furthermore, Akeakamai correctly responded to novel four-word instructions even though she had only been trained on two- and three-word sentences. Akeakamai does not always respond correctly but does so about 85 per cent of the time – far more than chance levels.

EVALUATION

The results from these studies seem to indicate that the dolphins were able to learn and respond to rules of grammar. They were sensitive to word order and responded accurately to novel sentences. There was also evidence of displacement in a number of ways. If asked to do something with a missing object the dolphins would search for it and then signal 'no' to indicate it was missing (this is spatial displacement since they were referring to an object that was not there). Also, if the dolphins were given an instruction about an object that was subsequently placed in the pool they were able to respond correctly (even though the instruction and object were separated in time or temporal displacement). The range of evidence provided by Herman and his colleagues seem to provide strong evidence of understanding of grammatical rules. However, there is no evidence yet that dolphins can produce sentences, and Pearce (1997) points out that language production is as important as language comprehension.

Is language 'human'?

I began this chapter by pointing out that there is a disagreement between two positions about animal language: the continuity school and the discontinuity school (Lewin, 1991). In other words there is a divide between those who believe that there is a quantitative difference in human language and animal language and those who believe that the difference is qualitative. There is a large range of studies of apes (chimpanzees, bonobos, gorillas and orang-utans) and other species, but despite analysing the same evidence there is still disagreement between the two schools.

Chomsky (1972) believes that there is a qualitative difference between human language and animal communication, and that language is unique to humans. He points to the similarities in grammar of all the human languages and suggests that there is an innate apparatus that generates a universal grammar. He calls this the language acquisition device (LAD) and suggests language is unique to humans because animals do not have a LAD. A slightly different view is taken by Vauclair (1990) who suggests that, although most complex cognitive activities show continuity between animals and humans, language does not. He believes that language in humans is the unique result of biological *and* social evolution that has resulted in an ability that is qualitatively different from that of animals.

In contrast Greenfield and Savage-Rumbaugh (1990) argue that animals do have the ability to use language. They agree that there is a genetic basis to human language but point out that language is likely to be influenced by a large number of genes and, since we have 99 per cent of our genes in common with chimpanzees, it is probable that chimpanzees and other primates will share much of the genetic basis for language. They also suggest there is a double standard in interpreting the linguistic abilities of young children and the use of symbols and signs in primates. Since children ultimately develop adult human language any combination of words or novel use of words is seen as evidence of a stage of language development. However, because primates do not develop the equivalent of an adult human language any novel use of words or any evidence of simple forms of grammar is questioned. After reviewing the evidence of language use in bonobos they believe the question is not 'Can animals learn language?' but 'What level of language development are animals capable of achieving?'

One problem in answering the question of whether language is unique to humans is that the concept of language is constantly evolving and, in refining the concept, philosophers and psychologists constantly pose new challenges for those trying to teach animals language. Shanker *et al.* (1999) note that 'ape language research has been marked by constantly shifting demands made by those who feel that only human beings can acquire language' (p. 24). They believe that it would be more productive to explore what the apes can do rather than searching for language skills that the apes have not mastered. Another problem of ape language research highlighted by Pearce (1997) is that absence of language learning may be caused by a lack of motivation rather than a lack of language ability or intellectual capacity. In other words the lack of language in animals may be because they *do not want* to communicate rather than they *cannot* communicate. A review of the studies on primates shows that those who had the most formal and uninteresting training (e.g. Nim) displayed the least productive and spontaneous language abilities. In contrast the primates who had the richest environments and who were given stimulating experiences (e.g. Kanzi) showed the most productive and spontaneous language skills. Shettleworth (1998) notes that Kanzi only developed his comprehension of language after intensive exposure to language and a lot of attention from adult humans, but such an experience is the norm for children.

The debate about animal language is interesting and complex and anyone who wishes to explore the question in more detail should compare the conclusions drawn by Terrace *et al.* (1979) in Key research summary: Article 3 (pp. 116–119) and Rumbaugh and Rumbaugh

(1994) in Key research summary: Article 4 (pp. 119–122). The contrast between them serves as a good illustration of the two positions on the question 'Is language uniquely human?' A more detailed analysis of these differences can be found by comparing the views of Wallman (1992) in *Aping Language* with those of Savage-Rumbaugh and Lewin (1994) in their book *Kanzi: The Ape at the Brink of the Human Mind*. A final point about the possibility of animals learning language is that there are reasons, other than purely scientific ones, why this debate is followed so closely. Shettleworth (1998) concludes: 'Given how closely the results of language-training projects bear on ideas about what makes us human, controversy about them is likely to continue' (p. 562).

If animals cannot use language then humans can still regard themselves as unique and separated from the rest of the animals. However, if animals can master language to a degree then we can no longer regard ourselves as separate, we are different in the level of ability only (quantitatively but not qualitatively different).

Summary

The debate about whether animals can use language is intense and remains unresolved, partly because of the complexity of language and partly because of methodological problems. Human language is a complex form of communication that can be distinguished from most animal communication by a number of key features such as displacement, productivity, prevarication and semanticity. A number of animals show *some* of these key features of language in their natural communication. For example, the dances of bees show displacement and refer to a food source that an individual bee found away from the hive some time ago. Vervets show semanticity and traditional transmission, and there is some evidence of prevarication in baboon troops. However, no natural system of communication shows *all* the key features of language.

An alternative way of studying language in animals is to try to teach them in controlled conditions, but early attempts to teach apes language failed because they concentrated on trying to teach speech. Later efforts using sign language seemed much more promising, and evidence shows that chimpanzees and gorillas are able to name objects and actions using signs. However, research by Terrace cast doubt on whether apes could use signs to spontaneously produce the complex sentences that are used in language. Alternative approaches using plastic tokens and lexigrams seem to indicate that chimpanzees can manipulate symbols to create sentences and that they respond appropriately to sentences created by their trainers. However, there is a possibility that these 'sentences' are merely complex behavioural sequences performed to gain some reward. Recent studies of bonobos indicate that if they are exposed to language at an early age they are able to learn to comprehend and produce sentences. Investigations of language comprehension in dolphins shows that they are able to respond to a novel sequence of words accurately; this suggests they are able to learn and respond to rules of grammar. However, although they show comprehension skills, they do not produce language. Despite the wealth of evidence to consider there is still a dispute about whether animals are capable of using language or whether language use is a purely human ability. Given the complexity (and sensitivity) of the subject it is a dispute that is likely to continue.

Review exercise

1 What major features of language are lacking in the natural communication of animals?
2 List four reasons from the study of Nim that cast doubt on the ability of primates to learn language (see also Key research summary: Article 3, pp. 116–119).
3 Do these four reasons apply to Kanzi (see also Key research summary: Article 4, pp. 119–122)?

Further reading

Pearce, J.M. (1997) *Animal Learning and Cognition* (2nd edn), Hove: Psychology Press. This book has an excellent chapter on animal communication and language which deals both with studies of natural animal language and with attempts to teach animals language.

Vauclair, J. (1996) *Animal Cognition*, Cambridge, Mass.: Harvard University Press. This book has a chapter on animal communication and human language that covers much of the same material as Pearce, but from a slightly different perspective.

6

Memory in non-human animals

Introduction

All animals need to satisfy certain basic needs in order to survive and for many animals some of these needs, such as finding the location of food and shelter, require some form of learning and memory. The study of memory in animals looks at how information that has been acquired at one time (learned) influences behaviour at a later time and traditionally investigates how information is stored and retrieved. Pearce (1997) points out that there are three main questions about memory in animals. How long can the information be retained? What type of information can be retained? How much information can be retained? The question of the type and amount of information that can be retained has been applied to two main areas: the use of memory in navigation and the use of memory in foraging for food.

Working memory and reference memory

In the study of human memory it has long been recognised that there seems to be a distinction between memories that last for a relatively short time and those which last for a long time period. These two types of memory are normally referred to as short-term memory and long-term memory, and there is an assumption that they work as different modules or memory systems (e.g. Atkinson and Shiffrin, 1968). A similar distinction is made in the study of animal memory, but typically the terms used are **working memory** and **reference memory** (Roberts, 1998). Working memory is used to refer to studies of short-term retention in animals and is defined as the memory for one specific trial since it contains the information needed for immediate purposes (Honig, 1978). The term 'working memory' is used to emphasise the fact that what is learned in one trial may not be relevant on the next, and does not necessarily refer to the same thing as Baddeley's (1986) concept of working memory in humans (which emphasises components of a short-term store). Reference memory is used to refer to long-term retention. In experimental studies it is the memory of the nature of the task over many trials, but in a broader sense it is all the memories the animal has of its behaviour and environment (Roberts, 1998).

To some extent the findings of the studies into working memory and reference memory in animals mirror those found in humans. Working memory is frequently studied with one or other of the variations of the *delayed matching to sample*. This involves presenting an animal with a stimulus and then removing it; then, after an interval (the retention interval), the animal is presented with the original stimulus and also with another. The animal is rewarded if it chooses the original stimulus. After training, this method can be used to investigate the effect of increasing the retention interval. Typically animals perform well on this task when the interval is short but as the interval increases the performance drops to chance level (Roberts, 1998). Working memory has also been studied using the **radial maze** (see pp. 88–89).

Reference memory in animals has been studied in a variety of ways and the findings typically suggest that, like long-term memory in humans, it has both a large capacity and is of long duration (Pearce, 1997). Theories of forgetting from reference memory originally concentrated on the processes of trace decay and interference, but there is increasing emphasis on the role of retrieval cues. This change of emphasis followed research that suggests that retrieval cues can reactivate memories in animals and has been extended to studies of both state-dependent memory (e.g. Overton, 1964) and context-dependent memory in animals (e.g. Gordon and Klein, 1994).

Memory in navigation (spatial memory)

In the discussion of navigational techniques in Cchapter 2 there was evidence that animals have a memory of the location of feeding sites, breeding grounds and nest/shelter sites which they use to navigate to the appropriate goal. For instance, the use of landmarks to orientate needs some form of representation of the landmarks, and the use of a compass technique (e.g. sun or magnetic compass) requires a representation of the relative positions of the starting point and goal point. Navigation therefore needs some form spatial memory or a representation of an area. This was shown in the van Beusekom (1948) study of digger wasps mentioned in Chapter 2 (see p. 20). He placed a circle of cones around the exit hole of a digger wasp's nest. Later, after the wasp had flown to and from the nest a number of times, he placed some cones in a square shape and others in an ellipse next to the circle. He found that the wasps sometimes made mistakes and went to the ellipse rather than the circle; but they did not go to the square. This suggests they had a representation of the circle that could be confused with the ellipse but not the square.

In a more controlled study Cartwright and Collett (1983) have shown that bees seem to use a representation of landmarks to locate food sources. They placed sugar solution in a room that was entirely white except for a black cylinder that was located at a fixed distance from the sugar. They filmed the behaviour of bees that found the sugar solution and found that on subsequent visits they began their search for the dish very close to the goal. When the black cylinder was removed the bees did not search near the goal. If a larger or smaller black cylinder was put in the original's place the bees searched in a different location (closer if the cylinder was smaller and further away if the cylinder was bigger). This suggests they used a representation of the cylinder to locate the food source and that they relied on a representation of the *apparent size* of the cylinder. When a larger cylinder was used its apparent size was only the same as the original if the bee was further away and if a smaller cylinder was used the apparent size was only the same as the original if the bee was closer.

In a review of spatial memory and navigation, Dyer (1996) discusses how honey-bees use information from landmarks and from the position of the sun in their foraging range (see Chapter 7, pp. 113–116). Bees only collect food for about ten days before they die, yet in that time they learn about spatial relationships in their environment and use information about familiar landmarks, the position of the sun and time of day both to find food sources and their hive. Dyer believes the evidence shows that bees do not use complex mechanisms such as cognitive maps (see pp. 91–96) but instead use a number of relatively simple means to learn to navigate in a very short time.

Spatial memory has been studied in laboratory based experiments using radial mazes (Figure 6.1). Radial mazes consist of a central platform with a series of identical arms branching out at equal angles (e.g. a four-arm maze would use a square central platform with an arm leaving the middle of each side). Olton and Samuelson (1976) tested a rat's spatial memory using an eight-arm radial maze. They placed food at the end of each arm and then put a hungry rat on the central platform and allowed it to explore. After a number of trials the rat typically went to each arm in succession. Although it did occasionally make mistakes, the rat's performance was far better than that expected by chance and showed a good spatial memory. Since the spatial memory was demonstrated in one trial it is also an example of working memory. The radial maze apparatus has been used to test the capacity of rats' spatial memory by analysing their performance on a seventeen-arm maze (Olton et al., 1977). The number of arms with food was gradually increased from two to seventeen. The rats made very few errors when they had up to eight choices and then showed an increasing number with each added choice. However, their performance even up to seventeen arms was much better than chance levels. Findings like these have led Roberts (1998) to conclude that the only limit to spatial memory may be the number of discriminal spatial locations in the environment (i.e. it is not limited by the memory capacity of the animal but the number of recognisable places).

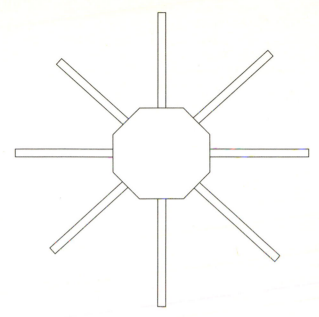

Figure 6.1 An eight-arm radial maze

Neural basis of spatial memory

There is a variety of evidence that suggests that spatial memory is linked to a brain structure called the *hippocampus*. The hippocampus in mammals is a fairly large structure in the limbic system, which is located beneath the cerebrum (see Figure 6.2). A similar structure to the hippocampus is also found in the brains of birds.

In mammals damage to this area causes disruption to spatial memory and poor performance in radial mazes (Roberts, 1998). This disruption seems to be selective and only affects some aspects of animal memory, particularly those involving spatial memory. For example, Morris *et al.* (1982) presented rats with two types of problem involving memory. In the first the rats swam towards two platforms which were visible above water but only one of which was safe. This was a visual discrimination task (i.e. it required the rats to learn the difference in appearance of the two different platforms), and the rats quickly learned to swim to the safe platform. The second task required the rats

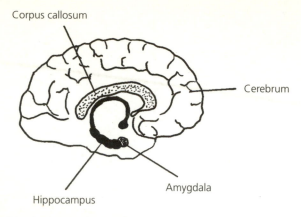

Figure 6.2 **The human brain, showing the hippocampus**

to swim to a platform hidden in opaque water. This involved the use of landmarks to find the platform and was a task that used spatial memory. Damage to the hippocampus did not affect the rats' ability to perform the first task but did impair their ability to do the second. Thus damage to the hippocampus does not seem to impair memory as a whole but does impair tasks requiring spatial memory.

A different type of evidence linking the hippocampus with spatial memory is provided by recordings of the activities of single cells within the structure. O'Keefe and Speakman (1987) monitored the activity of single cells in the hippocampus of rats as they moved around a radial maze. These cells are usually inactive but became active in particular locations: some cells were active in the goal arm of the maze whereas others were active in other arms. The cells were active even if familiar landmarks were removed. These recordings seem to reveal the existence of 'place' cells in the hippocampus.

There is also evidence that the hippocampus seems to be involved in spatial memory in birds. Removal of the structure impairs the ability of birds to use spatial memory. For example, Sherry and Vaccarino (1989) found that removal of the hippocampus in chickadees disrupted the bird's ability to recover food that they had stored (see food caching, pp. 98–99). These chickadees stored the food at the same rate as control birds, but unlike the controls they repeatedly visited empty sites and failed to visit sites where food was cached. Birds that store food have a greater need for spatial memory, and a comparative study of birds shows that these birds have a larger hippocampus than those of a similar species that do not store food (Sherry, 1992). This suggests that the enlarged hippocampus is an evolutionary adaptation to the demands made of spatial memory. There is also some evidence that the hippocampus in food-storing birds may undergo some seasonal variation in size, becoming larger during the seasons when food is stored and retrieved (Sherry and Duff, 1996).

There are alternative explanations for most of these studies; for example, the removal of the hippocampus in chickadees may affect their ability to navigate rather than their spatial memory. However, the accumulation of evidence from a variety of sources does tend to point to a link between spatial memory and the hippocampus. Several of these studies raise the issue of the ethics of research into animal behaviour since they involved the deliberate damage to the brains of either mammals or birds.

Cognitive maps

One theory of how animals find different objects and locations in their environments is that they form a complex representation, or a cognitive map, of their local habitat. The term was originally introduced by Tolman (1948) after performing a series of experiments with rats in mazes which he thought indicated that 'something like a field map of the environment gets established in the rat's brain'. Tolman believed that cognitive maps were representations of the routes and relationships between objects in the environment that enabled animals to make decisions about where to go. There are two sources of evidence for cognitive maps: the use of detours and the use of novel short cuts.

The use of detours to arrive at a goal point was shown in maze experiments on rats by Tolman and Honzik (1930). They designed a maze that had three routes from the start box to a food box and each route could be blocked (Figure 6.3). Route 1 was a direct route between the boxes and the other two were alternatives that varied in length. Tolman and Honzik found that the rats soon learned to use the direct route to the food box but if that was blocked they used the shorter alternative (route 2). If the path to the food box from both routes 1 and 2 were blocked near to the food box the rats returned to the start and took route 3. Thus the rats always took the shortest available route to

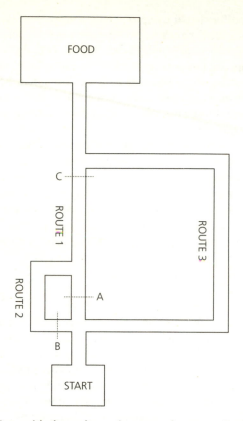

Figure 6.3 **Maze with three alternative routes between the start box and the food box**

If route 1 was blocked at A the rats took route 2. If both routes 1 and 2 were blocked at A and B the rats took route 3. If routes 1 and 2 were blocked at C the rats went back and used route 3.

Source: Based on E.C. Tolman and C.H. Honzik, '"Insight" in rats', copyright © 1930 The Regents of the University of California.

the food. This suggests that they used some spatial reasoning and had a cognitive map.

Another implication of cognitive maps is that they should enable animals to find alternative shorter routes between any two points in their locale, and a number of studies have shown that animals are able to take novel short cuts between goal points. For instance, Chapuis and Varlet (1987) tested the ability of Alsatian dogs to use short cuts.

They used a fixed start point and then walked the dogs to another point in the meadow (point A), where they were shown a piece of meat hidden in the grass before being led back to the start point (Figure 6.4). The dogs were then led in a different direction to another point in the meadow (point B) which was further away than point A. The dogs were shown another piece of meat hidden at point B before being led back to the start. All the dogs went first to the nearest piece of meat at point A, and in 96 per cent of the trials the dogs then went straight from point A to point B. They did not return to the start to take the route they had been shown but took a novel short cut.

Gould (1986) has reported a similar study with honey-bees that were taught to use two different feeding stations. If the food was removed from one site the bees flew directly to the second using a short cut rather than returning to the hive and flying to the second site on a familiar route. However, a number of studies have failed to replicate Gould's

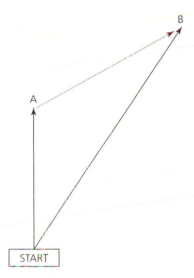

Figure 6.4 **Short cuts by dogs in natural surroundings**

Dogs were shown food at A then taken back to the start. They were then shown food at B and returned to the start again. On release they went first to A and then directly to B (shown by the dotted line).

Source: Based on N. Chapuis and C. Varlet, 'Shortcuts by dogs in natural surroundings', © 1987; reprinted by permission of The Experimental Psychology Society.

findings or have found that the bees only flew from one feeding station to another if there were prominent landmarks. For example, Dyer (1996) found that if he displaced bees as they were flying to feeding stations they compensated for the displacement when landmarks were visible and flew to the food source. However, when there were no familiar landmarks visible after displacement they did not find the food source (see Chapter 7, pp. 113–116). These findings suggest that the bees were using landmarks to navigate rather than a cognitive map.

Poucet (1993) has put forward a theory of how cognitive maps develop in animals. He suggests that cognitive maps are formed gradually by travel and exploration of the environment. He points out that there are two types of information that an animal might acquire about its locale: topological information and metric information. Topological information is the information about the relationships between objects, whereas metric information is more quantitative and refers to the distances and angles between objects. Poucet suggests that initially cognitive maps are based on topological information and are *location dependent* (that is, only certain points in the environment are recognised). As the animal explores its environment it learns more and more metric information and the cognitive map becomes more global and *location independent*. A location-independent cognitive map should allow animals to determine the direction from any point to any other point in its locale. Cognitive maps are therefore a type of reference memory since they deal with information that builds over time and are relatively long term.

EVALUATION

The notion of cognitive maps has been questioned by Bennett (1996) who claims that the term has been used to refer to a variety of spatial concepts. He points out that Tolman (1948) used the term to refer to a memory of landmarks which allows animals to take short cuts, whereas Gallistel (1994) uses the term to refer to any representation of space. He also notes that there are simpler explanations of novel short cutting, such as the use of familiar landmarks or the use of dead reckoning (see p. 23), and claims that no study of cognitive maps in any species has satisfactorily eliminated these alternatives. Dyer (1996) has investigated spatial memory in bees and points out that there are two explanations of how bees learn routes between the hive and a food source. One is that they learn to recognise a series of images (e.g. landmarks); the other is that they develop a cognitive map. He tested these two hypotheses by displacing bees which were flying towards feeding stations. Some displaced bees still had a view of landmarks but others did not. Only the bees that had a view of landmarks compensated for the displacement and found the feeding station. Dyer concludes that this suggests that bees do not have cognitive maps but use the simpler system of learning to navigate using a series of landmarks. Bennett believes the variety of definitions and the failure to eliminate simpler explanations of behaviour means that the notion of cognitive maps is no longer a useful concept for explaining the spatial behaviour of animals.

However, others (such as Roberts, 1998), disagree with this view. Roberts believes that there is evidence of cognitive maps from a variety of experiments that show that animals take detours and short cuts. He also points to evidence that some animals seem able to compute a path which requires the least distance among a number of food locations or other goals. For example, Menzel (1978) tested the ability of chimpanzees to remember the location of food in a familiar location. On each trial one chimpanzee was led around the location by the experimenter who hid 18 pieces of fruit. The chimpanzee was then led back to join five companions and a few minutes later they were all released. The chimpanzee that had accompanied the experimenter found over twelve pieces of fruit on average compared to less than 0.25 by the other chimpanzees. Furthermore the chimpanzee who had watched the food being hidden did not collect it in the order it had been placed but instead collected all the food in one part of the location before moving on to another. This suggests that the chimpanzee had a cognitive map of the area that enabled it to compute the shortest route to collect the food. An interesting recent development provides an alternative source of evidence that chimpanzees have cognitive maps. Kuhlmeier and Boysen (2002) used a scale model of a chimpanzee enclosure to demonstrate to chimpanzees the position of objects hidden in the enclosure. When taken to the enclosure the chimpanzees were able to find the hidden objects. This suggests that, not only did the chimpanzees have a cognitive map of the enclosure, but they also recognised the relationship between the model and the map.

The apparent contradictory nature of some of the evidence about cognitive maps may reflect the differences in the animals that have been tested. Acquiring and using cognitive maps requires an animal to process a lot of information, and it is possible that some animals are capable of doing so but others are not. For example, there seems to be little evidence that insects can use cognitive maps (e.g. Dyer, 1996), but some of the studies of mammals seem to suggest that they can (e.g. Menzel, 1978). Another reason for contradictory findings may be due to different interpretations of the concept of cognitive maps. In a detailed discussion of the topic Shettleworth (1998) notes: 'translating an intuitively appealing explanation of apparently intelligent behaviour in a way researchers agree on is seldom easy. When the results of behavioural tests cause theorists to revise ambiguous and slippery concepts, agreement can become almost impossible' (p. 317). It is probable that the 'slippery concept' of cognitive maps will be the subject of debate and disagreement for many years.

Progress exercise

Do humans have 'cognitive maps'?

Try drawing an accurate route between two points in your local area (choose a route that involves a few turns). Compare your map with an accurate scale map such as a good street map or OS map. Was your drawing accurate both in the directions and relative length of each segment?

If you were accurate can you think of an alternative explanation other than that you have a cognitive map?

Memory in foraging

In the discussion of memory in navigation there has been an assumption that the animal has learned a representation of a relatively fixed point. However, food sources are not fixed and they vary in quantity and position over time. An animal with a large range needs to forage for food and to remember where the best sources of food are at the present time. Roberts (1998) suggests that one of the main functions of spatial memory is to enable animals to remember the location of food sources in its habitat, and that this is critical for efficient **foraging**. The *optimal foraging theory* suggests that animals engage in behaviour that maximises the energy obtained from food for the time and energy spent foraging. Roberts points out that: 'If an animal is to travel through space and time to the best and nearest food sites while avoiding predators or competitors, it must have a memorial map of its habitat and of the locations of food, water, and predators within the habitat' (1998, p. 231). In an experimental demonstration of memory in foraging in rats Roberts (1992) placed cheese in the arms of a six-arm radial maze. The lumps of cheese in each arm varied in size but the same size lump was always placed in the same arm trial after trial. On the first few trials the rats entered the arms one after the other regardless of the size of the cheese, but later they started to visit each arm in a precise sequence. They went first to the arm with the biggest lump, then the next biggest, until finally visiting the arm with the smallest lump. This suggests they were using a spatial memory to locate the best sources of food (this is also an example of reference memory since the rat is using information from one trial to the next).

Many animals are faced with food sources that fluctuate and, if it is some time since a particular source was visited, it would make sense to respond to each source as if it contained the average of past food (Shettleworth, 1998). For example, if bird table A has food on it two-thirds of the time but bird table B has food on it only one-third of the time then, if they haven't visited either for a while, birds should visit table A first. A number of studies have confirmed that this is the case in both laboratory-based experiments (e.g. Devenport et al., 1997 with rats) and field studies (e.g. Devenport and Devenport, 1994 with ground squirrels and chipmunks). As with the Roberts study, this suggests a reference memory both of the locations of food sources and of the relative abundance of food at each source.

Foraging behaviour is not confined to locating food; animals need to forage for a variety of resources such as potential mates and nest/shelter sites. Sexual selection theory suggests that females choose the best mates rather than mating randomly (Shettleworth, 1998). One way of choosing is to inspect a number of potential mates and then return to the best. This would require a good memory of the quality and location of potential mates which avoids primacy and recency effects (i.e. remembering only the first and last potential mates that have been inspected). There is evidence for this theory in the wild in a number of animals where females have been observed to inspect and reject a male and then later re-inspect the male and accept him (Gibson and Langen, 1996). However, there is an alternative theory, the threshold theory, that accounts for mate selection and these findings and it doesn't rely on the use of memory so heavily. It suggests that the female will choose the first potential mate that meets a certain minimum standard or threshold. However, if no potential mate reaches this standard the female has to lower her threshold and may then accept a male that had previously been rejected. This may happen without a memory of the location of that or any other male in particular, just an impression of the general standard of males. The female may simply start a random search again, but with a lower standard.

There is evidence of the use of memory in the nest choice of the brown-headed cowbird. This is a brood parasite (like a cuckoo) and the female spends some time observing potential hosts building nests and then returns later to lay an egg in the chosen nests (Sherry *et al.*, 1993). Interestingly the hippocampus-like structure in the female brown-headed cowbird is larger than that of the males, presumably because she has a greater need for spatial memory since only she has to remember the location of the nest sites.

Food caching

Some animals avoid the problems of finding food in lean times of the year by storing food when it is plentiful. This storing of food for difficult times is called food caching and it requires a reference memory of the food stores to be useful. One familiar animal that stores food is the grey squirrel. Some field experiments have shown that if squirrels are removed from an area where they have buried some food and then returned some days later they are able to find the food accurately (MacDonald, 1997). In other words they do not seem to find food by chance but have a spatial memory of the food caches.

A large number of studies of memory in food caching have been carried out on two families of birds: tits and corvids (crow family). Field observations of marsh tits reveal that they regularly store food in a number of locations and then return some hours later to consume it. These observations do not show whether the tits find the stores by chance or whether they have a memory of the locations. Cowie *et al.*, (1981) found evidence of the latter. They initially located the marsh tits' caches of food using radioactive sunflower seeds. They replaced the radioactive seeds with normal seeds and also placed more seeds in similar hiding places nearby. The marsh tits ate the seeds from their original caches long before the other seeds, which suggests they were using spatial memory to relocate them rather than finding them in a random search of the area. One corvid that has been studied extensively for its food-caching behaviour is Clark's nutcracker (found in the USA). Nutcrackers bury thousands of pine seeds each autumn which they must find and eat to survive the winter. Studies of the nutcrackers suggest that they do not search randomly but dig for buried seed caches (Tomback, 1980).

Nutcrackers have also been studied in controlled artificial environments to test their spatial memory of food-cache location. In one study they were allowed to store seeds in any of a possible 180 locations and were then removed from the test site. When they were returned they consistently went to the seed caches at far above chance level (Kamil and Balda, 1983). Nutcrackers have been compared with other corvids that do not rely on stored food to the same degree and in all tests of spatial memory they seem superior. However, they do not perform better on other tests of memory, such as colour matching, and this suggests that their better performance in locating seed caches is based on a better spatial memory not a better memory generally (Shettleworth, 1998). Nutcrackers have a larger hippocampus than non-food-storing corvids and this adds further support for the hypothesis that spatial memory is linked to the hippocampus.

In a review of the behaviour of food-storing birds Sherry and Duff (1996) suggest that there is a preference ranking of the cues used to locate food. This preference starts with global cues, followed by local cues and then the features of the site. In other words the birds use finer and finer features of the environment until the cache has been found. Sherry and Duff point to two possible reasons for this. Firstly, the features of the cache sites are liable to change since the bird may store and retrieve the food in different seasons. The nutcracker can recover food that has been placed at a site seven months earlier. During this period there can be a lot of changes to the site caused by wind and rain and the site will differ because of seasonal changes (leaves falling in autumn, snow in winter, etc.). Secondly, the features of one site can be much like another and if birds relied on these alone they would have to search many areas in their habitat to find the cache sites.

Summary

There are two types of memory that are studied in animals: working memory and reference memory. Working memory is a relatively short-term memory that animals use to deal with information needed for immediate purposes, whereas reference memory is a relatively long-term memory that animals use to store information between trials or situations. Memory is important in navigation, and there is evidence from both field observations and laboratory experiments that animals have a spatial memory. Studies using radial mazes suggest that animals can store a lot of information about location. A variety of evidence links the ability to use spatial memory to the hippocampus region of the brain. One theory of how animals find the location of objects such as food or nests is that they have cognitive maps. Evidence for the existence of cognitive maps comes from studies that show that animals are able to use detours and novel short cuts. However, the term 'cognitive map' has been used in a variety of ways and there are alternative explanations of how animals use novel short cuts. It seems unlikely that animals such as insects have cognitive maps, but it is possible that some mammals do. Memory is important in foraging behaviour since it allows animals to remember the best locations of a resource. Studies of memory in foraging for food suggest that some animals are able store information about both the location and size of food sources. Animals may also use reference memory in foraging for mates and nests. One way of surviving periods when food is scarce is to store or hoard food. Studies of food-storing birds (known as food caching) show that they have a better spatial memory than similar species that do not engage in caching behaviour. Birds that use food caches also tend to have a relatively larger hippocampus than those that do not store food.

Briefly describe the following concepts and review the text to find an example of a study of each one.

Concept	Description	Example
Working memory		
Spatial memory		
Cognitive map		
Food caching		

Further reading

Roberts, W.A. (1998) *Principles of Animal Cognition*, Boston, Mass.: McGraw Hill. This book contains several chapters on animal memory, dealing with the topic very thoroughly.

Shettleworth, S.J. (1998) *Cognition, Evolution and Behavior*, New York: Oxford University Press. This book contains a very thorough account of memory in non-human animals and deals with methodological issues as well as research findings.

7

Study aids

IMPROVING YOUR ESSAY-WRITING SKILLS

At this point in the book you have acquired the knowledge necessary to tackle the exam yourself. Answering exam questions is a skill, and in this chapter we hope to help you improve this skill. A common mistake that some students make is not providing the kind of evidence the examiner is looking for. Another is failing to answer the question properly, despite providing lots of information. Typically, a grade C answer is accurate and reasonably constructed, but has limited detail and commentary. To lift such an answer to an A or B grade may require no more than fuller detail, better use of material and a coherent organisation. By studying the essays below, and the comments that follow, you can learn how to turn your grade C answers into grade A. Please note that marks given by the examiner in the practice essays should be used as a guide only and are not definitive. They represent the 'raw marks' which would be likely to be given to answers to AQA questions; that is, the marks the examiner would give to the examining board based on a total of 24 marks per question broken down into Skill A (description) and Skill B (evaluation). A table showing this scheme is in Appendix C of Paul Humphreys' title, *Exam Success in AQA (A) Psychology*. They may not be the marks given on the examination certificate ultimately received by the student because all examining boards are required to use a common standardised system called the

Uniform Mark Scale (UMS) which adjusts all raw scores to a single standard acceptable to all examining boards.

The essays are about the length a student would be able to write in 35–40 minutes (leaving you extra time for planning and checking). Each essay is followed by detailed comments about its strengths and weaknesses. The most common problems to look out for are:

- Students frequently fail to answer the actual question set, and present 'one they made earlier' (the *Blue Peter* answer).
- Many weak essays suffer from a lack of evaluation or commentary.
- On the other hand, sometimes students go too far in the other direction and their essays are all evaluation. Description is vital in demonstrating your knowledge and understanding of the selected topic.
- Don't write 'everything you know' in the hope that something will get credit. Excellence is displayed through selectivity, therefore improvements can often be made by *removing* material which is irrelevant to the question set.

For more ideas on how to write good essays you should consult *Exam Success in AQA (A) Psychology* (in this series) by Paul Humphreys.

Practice essay 1

Describe and evaluate two or more explanations of homing behaviour. (24 marks)

This essay title requires the candidate to make a decision about how many explanations to include. The title directs the candidate to provide at least two explanations (one explanation would therefore be marked as a 'partial performance'), but leaves the option to write about many explanations. This raises the issue of breadth and depth. Too many explanations would show evidence of breadth of knowledge but would cover each explanation at a superficial level; there would be a lack of depth. Restricting the answer to two explanations allows the candidate to discuss these explanations in depth, but the candidate would not demonstrate the breadth of their knowledge. One way of tackling this dilemma is to briefly point out that a range of explanations exist (breadth) and then to discuss two or three in detail (depth).

Homing behaviour is that which is directed at a specific goal – unlike navigation, which is to do with migration, possibly to a general area. Animals may use homing behaviour to find food or to return home, as exhibited by homing pigeons. Some journeys undertaken by animals may involve both migration and homing, and different strategies may be employed by the animal to negotiate long-distance travel (in the case of migration) and to seek shelter (which will involve homing behaviour).

Animals leaving their habitat have to return, they need to find food, water, etc. but need a means of locating their home again; homing is a technique they use to achieve this goal. Waterman (1989) says that an animal's habitat usually consists of a number of subhabitats and that an animal needs to be able to move about these to maximum effect. Psychologists have long been interested in investigating the types of mental processes at work when an animal engages in homing behaviour. This enables them to gain insight into the mechanisms animals use in their natural environment. A number of explanations have been used to enable psychologists to gain an understanding of the processes involved in homing behaviour.

There are a number of explanations that have been used to show how animals navigate home, such as the use of landmarks, the night sky and possibly magnetic fields that may guide the animal to the environment it seeks. McFarland (1999) identified three types of orientation that may be useful to navigation. These include pilot navigation, where familiar landmarks/features may be located; compass orientation, where a particular compass direction is located without the use of landmarks; and true navigation, which means that the animal can return to their habitat without using landmarks and regardless of compass direction. It is possible that pigeons that are able to home in on their loft no matter where they have been released use this third type of navigation.

One explanation of homing behaviour is that animals can use a compass of some sort to find home. A lot of animals forage for food and water during the day and could use the sun as a compass to let them know which direction to go. However, the sun moves from east to west during the day and animals would need an internal clock to use the sun. Clock-shifting experiments have tried to trick animals into believing it is the wrong time of day to see if it affects their ability to navigate. Experiments on pigeons show that if their clocks are shifted from mid-morning to mid-afternoon they fly off at 90 degrees from the correct direction. This only happens when it is sunny; on cloudy days clock shifting has no effect. This suggests they use a sun compass when they can see it. On cloudy days they may use a magnetic compass, and there is evidence that pigeons use magnetic fields to find north and south. Some researchers claim that birds can do this by detecting the angle of the magnetic field to the earth's surface. One problem with the use of a compass technique to find home is that compasses only point to a direction and unless the animal knows which direction to go this technique is useless. To use a compass you need a map, but there is little evidence that animals have mental maps.

Another explanation for homing behaviour is olfactory cues. These are smells and can be used by an animal to help them to home and navigate. This explanation suggests that animals learn a 'smell map' of their area. Salmon seem to use these cues to identify their home stream; pigeons, too, may use this as an aid to navigation. Ioale *et al.* (1990) used air currents containing the odour of benzaldehyde in pigeon lofts. The control group had normal air currents but the experimental group were misled and went in the opposite direction. Other studies have also shown that pigeon homing behaviour can be disrupted by false odour information (e.g. Benvenuti and Wallraff, 1985).

There is no doubt that animals use a wide variety of cues to enable them in homing and navigation, the use of a sun compass and olfactory cues being just two explanations out of a variety of possibilities. We cannot be absolutely sure exactly which techniques an animal is using, but psychologists have devised experiments to show us that certain techniques are occurring in the process of animal homing and navigation.

Examiner's comment

This essay indicates that the candidate has a good general knowledge of this area, but the essay lacks the focus and selectivity needed for a grade A or B. The candidate introduces a number of explanations of homing in the essay and then focuses on compass techniques and olfactory cues. There is some good description throughout the essay but, apart from the section on compass techniques, there is little evaluation

or commentary. This essay is likely to gain 13 marks: 8 for description and 5 for evaluation.

The paragraph on compass techniques contains a balance of both description and evaluation. The discussion of the use of the sun as a compass is clear and describes the evidence from clock-shifting experiments well, and the implications of this evidence are mentioned. The discussion of a magnetic compass is accurate but less detailed and would benefit from a brief mention of some experimental evidence. The last two sentences of the paragraph demonstrate that the candidate is aware of the major problem of compass explanations.

The discussion of the second explanation, olfactory cues, is very brief and suggests that the candidate was short of time (because of the long introduction). The description of the use of smell in homing was basic. Although the description of the experiment to distort air currents in pigeon lofts was mentioned the candidate does not use the information to evaluate the theory. The essay does fulfil the requirement to describe two or more explanations, but only one is evaluated well.

Practice essay 2

Describe and evaluate attempts which have been made to teach non-human animals to use language. (24 marks)

Given the wealth of studies that are relevant for this essay candidates will need to select the material carefully. A key word in the title is '*evaluate*'. Often answers to questions such as this one are a list of studies that show good descriptive skills but contain little evaluation. Candidates should try to *assess* the contribution of research to our understanding of animal language learning rather than merely describe it. Research evidence can be from studies and/or theories.

Candidate's answer

The debate over whether or not non-human animals can use language is one that has interested primatologists and psychologists for many years. Lewin (1991) said that there were two schools of thought on the issue. The continuity school sees the ability to use language as a continuum, with human language at one end of the scale and a lack of language ability at the other. No animal is capable of producing the same language as humans but some can show the major features of a language, though to a lesser degree than ourselves. The discontinuity school believes that language belongs to humans only and that this is what makes us unique compared to animals. They dismiss evidence gathered on non-human animals as deceptive, believing it to be a pattern of learned behaviour taught by humans to animals.

Before researchers can do work in this area, it is useful to look at the characteristics which are typical of language. Hockett (1959, 1960) described sixteen design characteristics of human language. These range from sounds, feedback, semanticity, learnability and sounds arranged in words and sentences. Some of these characteristics apply only to human language and may not be relevant to sign language. Brown (1973) sees semanticity, productivity and displacement as the three most important characteristics. Most animal communication occurs when other animals are present, but we are capable of displacement; that is, we can discuss objects, concepts and events that have not yet happened or that occurred in the past.

There have been many attempts by researchers to teach animals, particularly primates, language cognition but because primates do not possess the same vocal structure as us this causes problems over interpretation as much of the work has concentrated on teaching animals sign language or using symbols. Early attempts by Kellogg and Kellogg (1933) and Hayes and Hayes (1951) were failures due largely to their attempts to get chimps to speak, which they are not capable of doing. Later work by Gardner and Gardner (1969) tried a different approach with 'Washoe'. They used American Sign Language. Washoe was taught to sign, and after four years had an impressive range of signs and could respond to questions given by her trainer. Koko, a gorilla, was also trained by Patterson (1978) using both signs and spoken language. Koko could use signs creatively to refer to new objects and apparently had her own form of abuse.

However, Terrace (1979) found little evidence of language usage in Nim, another chimpanzee. Terrace believed that Nim was merely copying signs in return for a reward. Also the order in which Nim made signs was not consistent and he did not show an increase in the number of signs he linked together. When he did produce longer sequences of signs he usually repeated them: e.g. 'eat Nim eat Nim'. After he studied videos of Nim and Washoe Terrace concluded that chimpanzees could not use language.

Rambaugh and Savage-Rambaugh (1994) did a study with a pigmy chimpanzee called Kanzi, who was treated as a young child and allowed to roam freely and given stimulating experiences. He communicated with humans via a lexigram board and showed that he could both understand and use sentences. Much of his communication was spontaneous and not in response to cues from his trainers. His comprehension skills were believed to be close to that of a two and a half year old child but his sentence production skills were closer to those of a one and a half year old. Savage-Rumbaugh's work shows that, as in humans, there may be a sensitive period during which a primate may be more likely to acquire an understanding of language. The work with Kanzi suggests that animals can demonstrate many of the features of human language, but to a limited degree.

There have been many studies, which have investigated the possibility of non-human animals responding to and understanding human language. Terrace has raised serious questions about the ability of primates to use sign language, but Kanzi and others have shown that they can exhibit many features of language as outlined by Hockett. Such studies may enable us to find out more about how primates may or may not understand language and may help to understand language use in humans more.

Examiner's comment

This candidate demonstrates a good overall knowledge of this topic and has described most of the well-known studies of language learning in primates (although there is no mention of other types of animal). However, the candidate does not use their knowledge of the topic to best advantage and often mentions an idea or a study without using it to '*evaluate* attempts which have been made to teach non-human animals to use language'. This essay would probably gain 14: 9 for description and 5 for evaluation.

The first paragraph introduces the topic well and sets the debate in context. The second paragraph discusses the nature of language, which is useful since the answer to the question 'can animals learn language' hinges upon definition of the word 'language'. However, the candidate's efforts are to some extent wasted since the features mentioned here were not used later to evaluate attempts to teach animals language.

The paragraph on Washoe and Koko was descriptive only and no attempt was made to evaluate material. The candidate mentioned semanticity, productivity and displacement in the previous paragraph and could have used these features to evaluate the studies of Washoe and Koko. For example, does the finding that Washoe can name objects mean she shows semanticity? When Koko uses signs 'creatively' is she showing productivity?

The section on Project Nim both describes the findings well and is used to evaluate the studies of teaching apes language. However, this also could have been improved by referring back to a key feature of language, productivity, since Terrace argues that chimpanzees can be taught to name objects but they cannot use signs productively. In other words they can use signs but they cannot use grammar. The section on Kanzi is brief but summarises the main methods and findings well and is used to address the question. His use of language is compared to that of humans, and the idea of a sensitive period for language learning in both humans and primates is mentioned. The candidate also provides some evaluation of the study. The concluding paragraph summarises the topic well and helps to improve the candidate's mark for evaluation/commentary.

KEY RESEARCH SUMMARIES

Article 1: K. von Frisch (1974) Decoding the language of the bee. *Science*, 185, 663–668.

Aim

This article in *Science* was a translation of the speech von Frisch gave after receiving his Nobel Prize for Physiology in 1973. The aim of the article was to outline the communication system of honey-bees by summarising some of the findings from a number of the many experiments he had performed.

BASIC TECHNIQUE

Method: During the course of his studies von Frisch conducted a large number of experiments over a 60-year period, but they were mostly variations of a technique he devised at the beginning. He placed a small colony of bees in an observation hive, which is a hive with glass windows that enables observers to record the behaviour of bees on the honeycomb. He then placed a feeding bowl with sugar solution in it near to the hive and as bees visited the bowl he marked them with various coloured dots so that they could be identified (both in the hive and on any subsequent visits to the feeding station). He often put scented oils, such as peppermint or lavender, into the sugar solution to make each source distinctive. The distance between the feeding station and the hive varied from a few metres to several kilometres.

Results: Von Frisch found that bees returning to the hive after finding a good food source performed dances that recruited other bees to use the source. The type of dance varied according to the distance of the food source from the hive. If the source was within 50 metres the bees performed a round dance, which consists of circling right and left. Bees near to the dancer followed her and then left the hive to find the source. Von Frisch investigated the role of scent in locating food by using a peppermint-scented source. He placed other scented bowls in the same meadow, some also scented with peppermint but others with lavender, thyme, etc. He found that the bees visited the peppermint-scented bowls but not the other scented bowls.

After systematically changing the distance of the food source from the hive von Frisch found that when the source was more than 50 metres away the type of dance changed from a round dance to a waggle dance. During this dance the bee walks in a straight line before circling in one direction and then follows the same straight line before circling in the other direction (and thus performs a figure of eight movement). While walking along the straight portion of the dance the bee waggles her abdomen. As with the round dance this excites interest in other bees, which then follow the dancer. Von Frisch found that the straight portion of the dance provided detailed information about the location of the distant food sources. The duration of the tail wagging corresponded with the distance of the food source, the further the source the longer the wagging. The wagging lasted approximately 0.5 seconds when the source was at a distance of 200 metres and increased to 4 seconds as the distance increased to 4,500 metres. The angle of the straight part of the dance gave information about the direction of the food source. If the food source was in a straight line between the hive and the sun the straight part of the dance was vertical. If the food source was at an angle 40 degrees to the left of a line between the hive and the sun the straight part of the dance was performed at a 40 degree angle left of vertical. To test whether the other bees of the hive responded to this information appropriately von Frisch did two further experiments: the fan experiment and the step-by-step experiment.

FAN EXPERIMENT

Method: This was designed to test the ability of the worker bees to respond to communication about the direction of a food source. A few marked bees were trained to use a feeding station at 600 metres from the hive. Concentrated sugar solution with highly scented oil was then placed at the feeding station. When the foraging bees had found the scented food source von Frisch placed the same oil without food at seven other stations at a distance of 550 metres in a fan arrangement. The centre of the fan was in the same direction as the food source and the others were put in 15 degree steps either side. The number of unmarked bees visiting each of the seven test sites was recorded.

Results: In the 50 minutes following the introduction of the scented food source forty-two bees visited the scented plate that was in the same direction. Only twelve bees visited the other six scented plates, and the majority of those (seven) visited the plates either side of the correct direction.

STEP-BY-STEP EXPERIMENT

Method: This was designed to test the ability of the worker bees to respond to communication about the distance of a food source. In a procedure similar to the fan experiment marked bees were taught to visit a feeding station 2,000 metres from the hive and then a concentrated sugar solution with scented oil was placed there. Twelve other stations baited with the same oil without food were placed at various distances from the hive, but all were in the same direction as the food source. The distances ranged from 10 to 5,000 metres with two 50 metres either side of the original food source. The number of unmarked bees visiting each of the twelve test sites was recorded.

Results: No bees visited the scented plates that were very close to the hive or beyond 3,000 metres away. However, there was a sharp rise in the number of bees visiting a plate the closer to the original source it was: ninety-two bees visited the plate at 1,950 metres, with just twenty-eight visiting the plate at 2,050 metres.

Discussion

Von Frisch discovered that honey-bees use a system of communication that indicates the location of a food source with great accuracy. The waggle dance gives information about the distance and direction of a food source and other bees can use this information to find the source. The waggle dance showed that the bees used the sun as a compass and were able to use this to navigate to a source of food and then back to the hive. These studies also demonstrate the role of memory in foraging. It takes some time for bees to fly several kilometres, but a bee discovering a good source at this distance is able to communicate the direction and distance accurately to other bees. These bees use the information to fly to the source even though some time may have passed since they observed the dance.

Article 2: F.C. Dyer (1996) Spatial memory and navigation by honeybees on the scale of the foraging range. *Journal of Experimental Biology*, 199, 147–154.

Aim

The aim of this article is to explain how honey-bees find their way from the hive to distant feeding sites. It does this by reviewing research progress on questions about three aspects of honey-bee behaviour: (1) how do they learn about the spatial relationships between separated locations; (2) how do they learn to use the position of the sun; and (3) do bees learn to use the relationship between landmarks and the sun's position.

(1) How do honey-bees learn about the spatial relationships between separated locations?
Dyer notes that there is now a great deal of evidence that insects use information from visual landmarks in their foraging range to navigate. Some insects use landmarks to navigate over long distances, sometimes

up to several thousand metres. There is a question of how they use landmarks to navigate beyond the immediate location of the nest. In essence there are two possibilities: (a) they can learn routes by storing a series of images about the route (a route map), or (b) they can develop a 'large-scale metric map' (or cognitive map). The route map hypothesis assumes that the bees learn individual routes by storing a series of images about each one. Navigation consists of matching each of these images towards either a food source or the hive. This hypothesis predicts that bees can only move from one familiar route to another if they can see a landmark that matches a familiar image.

The cognitive map hypothesis suggests that the bees not only learn individual routes but also form a representation of each route in a common frame of reference, or a map. This would enable the bees to move from one familiar route to another by computing where it was on the map. This theory predicts that the bees should be able to take novel short cuts from one route to another because they could pinpoint where they were on the large-scale map. Evidence that bees could take novel short cuts was provided by Gould (1986), who believed that he had found indication of cognitive maps. However, Dyer found this explanation of bee navigation puzzling because the bee would have to construct a large-scale map from experience of each part over a long timescale. The bee would never be able to experience an overview of the whole terrain.

Dyer tested the two hypotheses by using a simple test: he displaced bees en route to feeding stations. After displacement some bees still had a view of the landmarks of the foraging route and these compensated for the displacement and headed directly for the food source. The other bees had no view of any familiar landmarks after displacement and on release did not fly to the food source. Dyer suggests this is because the bees had learned route maps based on familiar landmarks and not a cognitive map.

(2) *How do honey-bees learn to use the position of the sun?*
Bees do not always rely on familiar landmarks to navigate. Von Frisch (see Article 1, pp. 110–116) found that bees communicate the position of a good food source by using the waggle dance, which indicates the direction of the source with reference to the sun. This recruits new workers to the source who must navigate using the sun as a compass. However, this introduces a problem: the sun moves from east to west

during the day and the rate at which its position shifts varies (it is slower as it sets and rises, faster at midday). Dyer points out that many species, including honey-bees, seem to estimate the position of the sun even at times that they have never seen it. Thus bees that have been exposed to the sun in the afternoon seem to be able to estimate its position in the morning. It has been assumed that animals can do this by measuring the position of the sun and its rate of movement and, by using an endogenous time sense, then estimate the sun's course relative to landmarks surrounding the hive. There were three hypotheses of how bees do this: (a) they might interpolate between two known positions to estimate the current position; (b) they might extrapolate forwards from a point early in the day; or (c) they might extrapolate backwards from a position later in the day. There is some evidence to support each hypothesis.

Dyer tested the three hypotheses by studying the dances of incubator-reared bees on cloudy days to find how they estimated the position of the sun. The bees were initially exposed only to a small segment of the sun's course in late afternoon. They were tested in the morning and afternoon, but only on cloudy days when the sun was not visible and their dances were used as an indication of their estimate of the sun's position. The results were surprising and not explained by any of the original hypotheses. In the morning the bees estimated the sun's position to be approximately 180 degrees from the position they had learned in the afternoon. However, this position stayed more or less constant all morning until midday, then, instead of an incremental change in position, it changed by 180 degrees. The bees seem to estimate one morning and one afternoon position of the sun but not a gradual change in between. Dyer believes this suggests that the bees have an 'innate template' that indicates the approximate action of the sun and that this is modified with experience. If bees are exposed to the sun at several intervals in the day, or to the sun over a long time period, they learn that there is a gradual change in position. If, as in this experiment, they are exposed to one position of the sun for a brief period they assume it switches from one position to another.

(3) *Do bees learn to use the relationship between landmarks and the sun's position?*
Bees can use both landmarks and the sun to navigate independently, and there is some evidence that bees can use the position of landmarks to estimate the position of the sun. However, there is some doubt about

ANIMAL COGNITION

the degree of integration of the two systems. In a review of the evidence Dyer concludes that bees use both systems for pinpointing the position of a food source. Landmarks can be ambiguous, but a combination of a landmark and the sun's position (a compass direction) allows a location to be found precisely (bees find food near a landmark better on sunny days than cloudy days). However, Dyer has performed a number of studies that suggest that for navigation over a longer distance, such as returning to a hive, bees do not integrate information from landmarks and the sun's position.

Discussion

Dyer concludes that bees use a number of relatively simple mechanisms to learn navigational information rapidly. This is vital given their small nervous system and short life. However, the emphasis should be on the word 'relatively' since in the few days that bees collect food they learn a number of route maps, which seem to be based on a series of images of landmarks, and how to estimate the position the sun at any time of the day.

Article 3: H.S. Terrace, L.A. Petitto, R.J. Sanders and T.G. Bever (1979) Can an ape create a sentence? *Science*, **206, 891–900.**

Aim

The aim of this study was to answer the question set by the title 'Can an ape create a sentence?' This study came at the end of a decade of excitement following the publication of Gardner and Gardner's report on Washoe. Subsequent research on Sarah, Lana and Koko (see pp. 73–75) added to the impression that apes were capable of limited language and it seemed that the last barrier distinguishing humans from animals had been broken. Terrace and his colleagues carried out their own systematic study of sign-language learning on a chimpanzee named Nim and did a review of the other ape studies.

Method: From the age of two weeks Nim was raised by humans in a signing environment where *all* communication was in American Sign Language (ASL). He was taught to sign by moulding his hands to form the signs and by associating these signs with objects (i.e. in a similar

116

way to Washoe). Nim was not reinforced for specific combinations of signs, but his teachers used combinations of signs in a grammatically correct way. His teachers recorded each of his training sessions by whispering what he signed into a cassette recorder; they also noted whether the sign was spontaneous, prompted or moulded. A spontaneous sign was defined as one that was not moulded, prompted by the teacher or an imitation of the teacher. The criterion for sign acquisition was when three independent observers noted its use and when it occurred spontaneously on five successive days.

Results: Nim learned to use 125 signs that he combined firstly into two-sign combinations and then three or more sign combinations. More than 19,000 of these combinations were recorded and analysed. The majority of combinations were two-sign (11,845), followed by three-sign (4,294) and four-sign (1,587). Only 1,487 combinations were five signs or more and the average number of signs combined was 1.6. Analysis of the two-sign combination revealed a tendency to use some regular order of signs. For example, *more* + X was much more frequent (974 occasions) than X + *more* (124 occasions). Three-sign combinations showed less predictability and often seemed to be extended versions of the two-sign combinations. For example, the most common two-sign combination was 'play me' and the most common three-sign combination was 'play me Nim'. Often a three-sign combination was achieved by repeating one of the signs. For example, a common two-sign combination was 'eat Nim' and a common three-sign combination was 'eat Nim eat'. The tendency to repeat signs is also evident in the analysis of the four-sign combinations, the most common being 'eat drink eat drink' followed by 'eat Nim eat Nim'.

The average length of Nim's sign combinations was recorded each month between the ages of 26 and 45 months. These were compared to children's mean length of utterance (MLU), which is a measure of the length of their sentences. During the 19-month period there was no increase in the average length of Nim's sign combinations, and it stayed at approximately 1.6. However, during a similar period children show a rapid increase in MLU from 1.6 to over 4.

Analysis of videotape transcripts of Nim and his teachers signing revealed other differences between Nim and children. Studies of children show that they take turns in communication from an early age and that this is a precursor to the turn-taking we use in conversation.

Nim did not do this and would frequently sign at the same time as his trainers. Children are spontaneous in their language production and will frequently initiate conversation. Nim, on the other hand, seemed to respond to signs from his teachers and often appeared merely to imitate the signs made by the teacher. Thus what seems to be a sentence can be a sequence prompted by cues from the trainer. Terrace *et al.* reviewed film evidence from other studies and found that the other apes were also imitating their trainers or were responding to cues from them. They claim that, when analysed carefully, the famous sequence of Washoe signing 'baby in my drink' shows that she was responding to the trainer pointing to each of the objects in turn. An analysis of any sequence of signs on film of either Washoe or Koko showed that they were either prompted or were mirroring the signs made by the trainer.

Discussion

Terrace *et al.* concluded that the evidence from Project Nim and other studies of ape language learning suggested that apes could use symbols but that they could not combine these symbols to create new meanings. They noted that there was some regularity in the production of two-sign combinations but that it was unwarranted to conclude that these were primitive sentences. Beyond two-sign combinations there was no evidence of consistent structure, and many signs were repeated. The average number of signs in combinations did not increase significantly during the study and remained about 1.6. In contrast the MLU in children over a similar period tends to increase consistently. Furthermore, the way in which Nim used the signs differed from that of a child. Nim interrupted his teachers (there was no turn-taking as there is in conversation with a child) and imitated his teachers (rather than producing signs spontaneously). Terrace *et al.* believed that all evidence that an ape could create a sentence could be explained by simpler processes, concluding: 'Apes can learn many isolated symbols . . . but they show no unequivocal evidence of mastering the conversational, semantic, or syntactic organisation of language' (p. 901).

This report had a huge impact on the study of language learning in apes and for a while threatened to put an end to any further research. Savage-Rumbaugh has noted that following this article the study of ape language became a 'non science' and that funding for new projects stopped (Savage-Rumbaugh and Lewin, 1994). She points out that it was as if Terrace had *proved* that apes could not learn a language and therefore the subject was closed since any evidence to the contrary was not accepted by journals. The problem with this is that no one study can prove or disprove such a hypothesis – all that can be concluded is that Nim did not seem to learn sign language. However, it does not mean that no ape can ever learn by any method.

Article 4: D.M. Rumbaugh and E.S. Savage-Rumbaugh (1994) Language in comparative perspective. In N.J. Mackintosh (ed.), *Animal Learning and Cognition*. London: Academic Press.

Aim

Some fifteen years after the Terrace article Rumbaugh and Savage-Rumbaugh presented the alternative point of view of ape language learning. Their article is presented in six parts, the first being a discussion about the nature of language and the second a summary of teaching apes language up to Project Nim. This section ends by describing how the Terrace *et al.* (1979) article did not generate discussion but stifled most research and debate in this area for some years. This summary focuses on the next three sections of the article which discuss the research that survived following the Terrace *et al.* article. It concentrates on three apes: two common chimpanzees (Sherman and Austin) and one bonobo (Kanzi).

Method: Sherman, Austin and Kanzi used a system that was originally developed for Project Lana, called lexigrams, to communicate. Lexigrams are distinctive geometric patterns arranged on a large keyboard. Each lexigram key was linked to a computer and was used to represent a word. Communication could be two way: the trainers could make requests or ask questions using the lexigrams and the apes could use the lexigrams to either answer or to ask for things. To create a sentence the animals had to press the lexigrams in the correct sequence. The sequences produced using the lexigrams could be recorded accurately and objectively (unlike sign language, which has to be interpreted).

The other common element in the studies is that they concentrate on comprehension rather than production of language.

SHERMAN AND AUSTIN

The studies of Sherman and Austin focused on categorisation skills and semantics, since unless words (or in this case lexigrams) have meaning they cannot be used to communicate. They were trained in a number of stages. In the first stage they were given three tools and three food items and were required to place them in one of two bins, food in one and tools in the other. Thus the chimpanzees were taught to classify objects. In the second stage they were required to classify photographs of either food or tools. The final stage of training involved the learning of two new lexigrams, one for 'food' and the other for 'tool'. They were then presented with a number of lexigrams that they had learned previously which represented a variety of tools and food. The test was to find out whether they could categorise these correctly. The results showed that they could do so accurately and only one error was made by the two chimps. (Sherman categorised a *sponge* as a food, but Rumbaugh and Savage-Rumbaugh point out that he liked to suck juice from a sponge so from his perspective perhaps it was not a mistake!)

Sherman and Austin were also tested for their ability to use lexigrams to indicate a future course of action. They were shown a number of food and drink items in one room and then taken to another room that contained the lexigram keyboard. They were then required to choose one of the items they had seen. Finally they were taken back to the room with the food and drink and if they picked the item they had declared on the lexigram they were allowed to have it. There was approximately a 90 per cent concurrence between the lexigram and item choice.

KANZI

Initially no attempt was made to train Kanzi because the subject of the study was his adoptive mother Matata. Kanzi was merely an observer of her training sessions until he was two and a half years old when they had to be separated for a while. This was when it was decided to start training Kanzi. However, it was found that Kanzi already knew how to use the lexigram keyboard to name objects, to request things and to announce his actions (Matata in contrast had not learned to use the keyboard).

Savage-Rumbaugh decided to change the way that Kanzi was trained. Unlike previous ape language studies Kanzi was not formally trained but was immersed in a rich social life where people spoke to him and used the lexigram keyboard at the same time. Kanzi was told about all the events and activities that were to happen and people spoke to him as if he could understand. He was studied in a laboratory situated in a 55-acre forest, and on regular walks in the forest he was encouraged to name various sites. If he pointed to a lexigram that represented one of the sites he was taken there. Importantly, although Kanzi was encouraged to use the lexigram keyboard he was not denied objects or activities if he failed to use it or if he used it incorrectly.

By the age of five and a half Kanzi had learned to use 149 lexigrams on the keyboard and showed evidence of understanding human speech. Tests revealed that he not only understood the meaning of individual words but of sentences as well. Kanzi's language skills were compared directly to those of a human child, Alia, aged two and a half. Alia's mother was one of Kanzi's trainers and she used the lexigram keyboard with both Kanzi and Alia. Their ability to understand language was then tested in a series of carefully controlled tests. In the first series of tests Kanzi and Alia were given seven different types of sentences that tested their ability to understand instructions. For example, one type was 'Give object X and object Y to animate A' (e.g. Give the peas and beans to Kelly) while another was 'Go to location X and get object Y' (e.g. Go to the microwave and get the tomato). All of the sentences were novel to both Kanzi and Alia. The results showed that Kanzi followed the instructions correctly 74 per cent of the time and Alia was correct 66 per cent of the time.

A second series of tests were designed to test comprehension of different sentences that involved some kind of reversal. For example, 'Take the potato outdoors' versus 'Go outdoors and get the potato', or 'Put the doggie in the refrigerator' versus 'Get the dog that is in the refrigerator'. These tests showed that both Kanzi and Alia were sensitive to word order, with Kanzi being correct in 81 per cent of the trials and Alia correct in 64 per cent. Kanzi's ability to *comprehend* language therefore seems comparable to that of a human child of two and a half.

Further tests of Kanzi's ability to *produce* language revealed that it was comparable to that of a human child of one and a half. However, unlike many previous studies of language learning in apes, Kanzi usually produced sequences on his lexigram keyboard spontaneously and did not merely respond to his trainers. Rumbaugh and Savage-Rumbaugh claim that many of the sequences were declarations of what he was about to do.

Discussion

Rumbaugh and Savage-Rumbaugh believe that the success in teaching apes in their programmes is due to a number of fundamental differences to the early research. Firstly, with all three apes (Sherman, Austin and Kanzi) they concentrated on training comprehension skills rather than production skills, since they claim that this is the only way language can be learned. Human parents do not teach their children to speak then understand language, rather they talk to the child about the world around them. The child listens and begins to understand language, but the production of language, or speech, comes later. Rumbaugh and Savage-Rumbaugh point out that the early studies of apes concentrated on the production of language without the initial emphasis on comprehension. Imagine a person who is taught a hundred words of Russian and then told to converse with a Russian person. When they failed would the conclusion be that the person could never learn a foreign language or that they had to learn to understand some Russian before they could speak Russian?

The second major difference, which began with Project Kanzi, was to study language acquisition in apes that were brought up in a 'language-saturated environment' from a very early age. This approach treats the apes more like children in that they are exposed to an environment rich in language, though they are not formally taught to use signs in a particular way. The success of Project Kanzi, in which he first developed an understanding of language before learning to use language, has been replicated with other apes. Two of the apes, Mulika and Panbanisha, are bonobos and the other, Panzee, is a common chimpanzee. Comparisons of the two species suggest that bonobos learn to comprehend and produce language more readily than common chimpanzees. Rumbaugh and Savage-Rumbaugh believe these studies suggest that there is a sensitive age for language acquisition in apes.

Up-to-date progress on Kanzi, Mulika, Panbanisha and Panzee can be found on the Language Research Centre's website: www.gsu.edu/~wwwlrc/biographies.html

Glossary

animat   An animal robot which is designed to mimic animal behaviour.

anthropomorphism   Ascribing human characteristics to animals. Describing and explaining animal behaviour in human terms.

anthropocentrism   The tendency to view the world from a human point of view.

circadian rhythm   A rhythm that has an approximate 24-hour cycle.

circannual rhythm   A rhythm that has an approximately yearly cycle.

clock-shifting   The shifting of the circadian rhythm of an animal to test the relationship between an endogenous clock and use of the sun compass.

cognition   The mental processes that allow animals to process information.

cognitive maps   A mental representation of the environment that establishes the location of objects relative to each other.

communication   The transmission of information that is designed to influence the behaviour of the receiver.

compass orientation   Navigation in one particular direction.

dead reckoning   The ability to return directly to a starting point no matter how complex the outward route.

displacement   The ability to refer to something which is removed in space or time.

endogenous factors   Factors that come from within. These are in-built or biological factors.

food caching   Storing food for use at a later time.

foraging   The search for resources such as food.

functionally referential signals   Signals that are made in response to a specific object or event that elicit a consistent behaviour in the receiver.

grammar   Rules of language which govern word order and sentence construction.

homing   The ability to return to a shelter or nest after leaving it.

language   A sophisticated form of communication which allows the user to produce and understand a large number of messages.

Lloyd Morgan's canon   If a behaviour can be explained using a simple process and a complicated process then the simple explanation should be accepted.

migration   The seasonal, orientated and relatively long-distance movements of animals.

migratory restlessness   The agitated behaviour of birds which are caged during their migration period.

multimodal communication   Simultaneous communication through two or more channels.

navigation   The use of cues to orientate from one location to another.

pheromone   A chemical that transmits a signal from one animal to another.

pilotage   Use of features from the environment to orientate or navigate.

productivity   The use of rules (grammar) to produce novel sentences.

prevarication   The use of language to communicate false information.

radial maze   A maze with a number of identical arms emanating from a central platform.

reference memory   A relatively long-term memory that is used to deal with information across trials or situations.

round dance   The left and right movement that honey-bees use to indicate a local food source.

semanticity   The use of symbols to represent objects or actions.

signal   Coded information that is sent by an animal.

spatial memory   A memory of the position of objects.

true navigation   The ability to find a goal point from any starting point regardless of direction or distance.

waggle dance   A dance performed by honey-bees to indicate the direction and distance of a food source.

working memory   A relatively short-term memory that is used to deal with information needed for immediate purposes.

Bibliography

Able, K.P. (1980) Mechanisms of orientation, navigation, and homing. In S.A. Gauthreaux (ed.), *Animal Migration, Orientation, and Navigation*. New York: Academic Press.

Able, K.P. (1996a) Large-scale navigation. *Journal of Experimental Biology*, 199, 1–2.

Able, K.P. (1996b) The debate over olfactory navigation by homing pigeons. *Journal of Experimental Biology*, 199, 121–124.

Alcock, J. (1993) *Animal Behaviour* (5th edn). Sunderland, Mass.: Sinauer Associates.

Allen, C. and Hauser, M. (1996) Concept attribution in nonhuman animals. In M. Bekoff and D. Jamieson (eds), *Readings in Animal Cognition*. Cambridge, Mass.: The MIT Press.

Altmann, S.A. (1962) A field study of the sociobiology of rhesus monkeys *Macaca mulatta*. *Annals of the New York Academy of Science*, 102, 338–435.

Atkinson, R.C. and Shiffrin, R.M. (1968) Human memory: A proposed system and its control processes. In K.W. Spence and J.T. Spence (eds), *The Psychology of Learning and Motivation: Advances in Research and Theory*. New York: Academic Press.

Baddeley, A.D. (1986) *Working Memory*. Oxford: Oxford University Press.

Baker, R.R. (1978) *The Evolutionary Ecology of Animal Migration*. New York: Holmes and Meier Publishers.

Baker, R.R. (1980) *The Mystery of Migration*. London: Macdonald Futura Books.

Baker, R.R. (1982) *Migration. Paths through time and space*. London: Hodder and Stoughton.

Bastion, J. (1967) The transmission of arbitrary environmental information between bottlenosed dolphins. In R.G. Busnel (ed.), *Animal Sonar Systems* (Vol. 2). Jouy-en-Josas, France: Laboratoire de Physiologie Acoustique.

Bekoff, M. and Jamieson, D. (1996) *Readings in Animal Cognition*. Cambridge, Mass.: The MIT Press.

Bellrose, F.C. (1958) Celestial orientation in wild mallards. *Bird Banding*, 29, 75–90.

Bennett, A.T.D. (1996) Do animals have cognitive maps? *Journal of Experimental Biology*, 199, 219–224.

Benvenuti, S. and Wallraff, H.G. (1985) Pigeon navigation: site simulation by means of atmospheric odours. *Journal of Comparative Physiology*, 156, 737–746.

Berthold, P. (1998) Spatiotemporal aspects of avian long-distance migration. In S. Healy (ed.), *Spatial Representation in Animals*. New York: Oxford University Press.

Bonner, J.T. (1969) Hormones in social amoebae and mammals. *Scientific American*, 220, 78–87 (June).

Bookman, M.A. (1978) Sensitivity of the homing pigeon to an earth-strength magnetic field. In K. Schmidt-Koenig and W.T. Keeton (eds), *Animal Migration, Navigation and Homing*. New York: Springer-Verlag.

Braithewaite, V.A. (1998) Spatial memory, landmark use and orientation in fish. In S. Healy (ed.), *Spatial Representation in Animals*. New York: Oxford University Press.

Bright, M. (1984) *Animal Language*. London: BBC Publications.

Brower, L.P. (1996) Monarch butterfly orientation: Missing pieces of a magnificent puzzle. *Journal of Experimental Biology*, 199, 93–103.

Brown, R. (1973) *A First Language: The early stages*. Cambridge, Mass.: Harvard University Press.

Byrne, R. and Whiten, A. (1987) The thinking primate's guide to deception. *New Scientist*, 3 December, 54–57.

Byrne, R. and Whiten, A. (1988) *Machiavellian Intelligence: Social expertise and the evolution of intellect in monkeys, apes and humans*. Oxford: Clarendon Press.

Cartwright, B.A. and Collett, T.S. (1983) Landmark learning in bees. *Journal of Comparative Physiology A*, 151, 521–543.

Cartwright, J. (2002) *Determinants of Animal Behaviour*. Hove, UK: Routledge.

Catchpole, C. (1984) Song is a serenade for the warblers. In G. Ferry (ed.), *The Understanding of Animals*. Oxford: Blackwell.

Chapuis, N. and Varlet, C. (1987) Shortcuts by dogs in natural surroundings. *The Quarterly Journal of Experimental Psychology*, 39, 49–64.

Cheng, K. (1986) A purely geometric module in the rat's spatial representation. *Cognition*, 23, 149–178.

Cheng, K. (1994) The determination of direction in landmark-based spatial search in pigeons: A further test of the vector sum model. *Animal Learning and Behaviour*, 22, 291–301.

Cheng, K. and Spetch, M.L. (1998) Mechanisms of landmark use in mammals and birds. In S. Healy (ed), *Spatial Representation in Animals*. New York: Oxford University Press.

Chomsky, N. (1972) *Language and Mind*. New York: Harcourt Brace Jovanovich.

Clutton-Brock, T.H. and Albon, S.D. (1979) The roaring of red deer and the evolution of honest advertisement. *Behaviour*, 69, 145–170.

Collett, T.S., Cartwright, B.A. and Smith, B.A. (1986) Landmark learning and visuo-spatial memories in gerbils. *Journal of Comparative Physiology A*, 158, 835–851.

Cowie, R.J., Krebs, J.R. and Sherry, D.F. (1981) Food storing by marsh tits. *Animal Behaviour*, 29, 1252–1259.

Dawkins, M.S. (1995) *Unravelling Animal Behaviour* (2nd edn). Harlow: Longman.

Dawkins, R. and Krebs, J.R. (1978) Animal signals: information or manipulation? In J.R. Krebs and N.B. Davies, *Behavioural Ecology, an Evolutionary Approach*. Oxford: Blackwell Scientific Publications.

Devenport, L.D. and Devenport, J.A. (1994) Time-dependent averaging of foraging information in least chipmunks and golden-mantled ground squirrels. *Animal Behaviour*, 47, 787–802.

Devenport, L., Hill, T., Wilson M. and Ogden, E. (1997) Tracking and averaging in variable environments: A transition rule. *Journal of Experimental Psychology: Animal Behaviour Processes*, 23, 450–460.

Dyer, F.C. (1991) Bees acquire route-based memories but not cognitive maps in a familiar landscape. *Animal Behaviour*, 39, 17–41.

Dyer, F.C. (1996) Spatial memory and navigation by honeybees on the scale of the foraging range. *Journal of Experimental Biology*, 199, 147–154.

Emlen, S.T. (1972) The ontogenetic development of orientation capabilities. In S.R. Galler, K. Schmidt-Koenig, G.J. Jacobs and R.E. Belleville (eds), *Animal Orientation and Navigation*. Washington, D.C.: US Govt Printing Office.

Emlen, S.T. (1975) The stellar-orientation system of a migratory bird. *Scientific American*, 233, 102–111.

Evans, C.S. (1997) Referential signals. *Perspectives in Ethology*, 12, 99–143.

Evans, C.S. and Marler, P. (1995) Language and animal communication: Parallels and contrasts. In H.L. Roitblat and J.A. Meyer (eds), *Comparative Approaches to Cognitive Science*. Cambridge, Mass.: MIT Press.

Fisher, J.A. (1996) The myth of anthropomorphism. In M. Bekoff and D. Jamieson (eds), *Readings in Animal Cognition*. Cambridge, Mass.: The MIT Press.

Fouts, R.S., Hirsch, A.D. and Fouts, D.H. (1982) Cultural transmission of a human language in a chimpanzee mother–infant relationship. In H.E. Fitzgerald, J.A. Mullins, and P. Gage (eds), *Child Nurturance: III, Studies of development in nonhuman primates*. New York: Plenum.

Gallistel, C.R. (1994) Space and time. In N.J. Mackintosh (ed.), *Animal Learning and Cognition*. London: Academic Press.

Gardner, R.A. and Gardner, B.T. (1969) Teaching sign language to a chimpanzee. *Science*, 165, 664–672.

Gibson, R.M. and Langen, T.A. (1996) How do animals choose their mates? *Trends in Ecology and Evolution*, 11, 468–470.

Gordon, W.C. and Klein, R.L. (1994) Animal memory: The effects of context change on retention performance. In N.J. Mackintosh (ed.), *Animal Learning and Cognition*. London: Academic Press.

Gould, J.L. (1986) The locale map of honey bees: Do insects have cognitive maps? *Science*, 232, 861–863.

Gould, J.L. and Gould, C.G. (1988) *The Honey Bee*. New York: Scientific American Library.

Greenfield, P.M. and Savage-Rumbaugh, E.S. (1990) Grammatical combination in *Pan paniscus*: process of learning and invention in the evolution and development of language. In S.T. Parker and K.R. Gibson (eds), *'Language' and Intelligence in Monkeys and Apes: Comparative Developmental Perspectives*. New York: Cambridge University Press.

Grier, J.W. and Burke, T. (1992) *Biology of Animal Behaviour*. Dubuque, Ia.: Wm. C. Brown Communications.

Griffin, D.R. (1958) *Listening in the Dark*. New Haven, Conn.: Yale University Press.

Griffin, D.R. (1984) *Animal Thinking*. Cambridge, Mass.: Harvard University Press.

Gwinner, E. (1986) Internal rhythms in bird migration. *Scientific American*, 254, 84–92.

Gwinner, E. (1996) Circadian and circannual programmes in avian migration. *Journal of Experimental Biology*, 199, 39–48.

Hasler, A.D., Scholz, A.T. and Horrall, R.M. (1978) Olfactory imprinting and homing in salmon. *American Scientist*, 66, 347–355.

Hayes, K.H. and Hayes, C. (1951) Intellectual development of a house-raised chimpanzee. *Proceedings of the American Philosophical Society*, 95, 105–109.

Healy, S. (1998) *Spatial Representation in Animals*. New York: Oxford University Press.

Herman, L.M. and Morrel-Samuels, P. (1996) Knowledge acquisition and asymmetry between language comprehension and production: Dolphins and apes as general models for animals. In M. Bekoff and D. Jamieson (eds), *Readings in Animal Cognition*. Cambridge, Mass.: The MIT Press.

Herman, L.M., Pack, A.A. and Morrel-Samuels, P. (1993) Representational and conceptual skills of dolphins. In L.M. Roitblat, L.M. Herman and P.E. Nachtigall (eds), *Language and Communication: Comparative perspectives*. Hillsdale, N.J.: Lawrence Erlbaum Associates.

Herman, L.M., Richards, D.G. and Wolz, J.P. (1984) Comprehension of sentences by bottlenosed dolphins. *Cognition*, 16, 129–219.

Hinde, R.A. and Rowell, T.E. (1962) Communication by postures and facial expressions in the rhesus monkey (*Macaca mulatta*). *Proceedings of the Zoological Society of London*, 138, 1–21.

Hockett, P.C. (1959) Animal languages and human language. *Human Biology*, 31, 32–39.

Hockett, P.C. (1960) The origins of speech. *Scientific American*, 203, 89–96.

Honig, W.K. (1978) On the conceptual nature of cognitive terms: An initial essay. In S.H. Hulse, H. Fowler and W.K. Honig (eds), *Cognitive Processes in Animal Behaviour*. Hillsdale, N.J.: Erlbaum.

Humphrey, N.K. (1976) The social function of intellect. In P.P.G. Bateson and R.A. Hinde (eds), *Growing Points in Ethology*. Cambridge: Cambridge University Press.

Huxley, J.S. (1914) The courtship of the great crested grebe. *Proceedings of the Zoological Society of London*, 2, 491–562.

Ioale, P., Nozzolini, M. and Papi, E. (1990) Homing pigeons do extract navigational information from olfactory stimuli. *Behavioural Ecology Sociobiology*, 26, 301–305.

Kamil, A.C. and Balda, R.P. (1983) Cache recovery and spatial memory in Clark's nutcrackers *(Nucifraga columbiana)*. *Journal of Experimental Psychology: Animal Behaviour Processes*, 11, 95–111.

Keeton, W.G. (1969) Orientation by pigeons: Is the sun necessary? *Science*, 165, 922–928.

Keeton, W.G. (1974) The orientational and navigational basis of homing in birds. *Advances in the Study of Behaviour*, 5, 47–132.

Kellogg, W.N. and Kellogg, L.A. (1933) *The Ape and the Child*. New York: McGraw-Hill.

Kennedy, J.J. (1992) *The New Anthropomorphism*. Cambridge: Cambridge University Press.

Köhler, W. (1925) *The Mentality of Apes*. New York: Harcourt Brace.

Kramer, G. (1952) Experiments on bird orientation. *Ibis*, 94, 265–285.

Krebs, J.R. (1984) The song of the great tit says 'Keep Out'. In G. Ferry (ed.), *The Understanding of Animals*. Oxford: Blackwell.

Krebs, J.R. and Davies, N.B. (1993) *An Introduction to Behavioural Ecology* (3rd edn). Oxford: Blackwell Scientific Publications.

Kreithen, M.L. and Keeton, W.T. (1974) Attempts to condition homing pigeons to magnetic stimuli. *Journal of Comparative Physiology*, 91, 355–362.

Kuhlmeier, V.A. and Boysen, S.T. (2002) Chimpanzees recognize spatial and object correspondences between a scale model and its referent. *Psychological Science*, 13, 60–63.

Lewin, R. (1991) Look who's talking now. *New Scientist*, 27 April, 48–52.

Lissman, H.W. (1963) Electric location by fishes. *Scientific American*, 207, 50–59.

Lohmann, K.J. and Lohmann, C.M.F. (1996) Orientation and open-sea navigation in sea turtles. *Journal of Experimental Biology*, 199, 73–81.

Lorenz, K.Z. (1952) *King Solomon's Ring* (M.K. Wilson, trans.). New York: Thomas Y. Crowell.

MacDonald, I.M.V. (1997) Field experiments on duration and precision of grey and red squirrel spatial memory. *Animal Behaviour*, 54, 879–891.

McFarland, D. (1999) *Animal Behaviour* (3rd edn). Harlow: Longman.

Mackintosh, N.J. (1994) *Animal Learning and Cognition*. London: Academic Press.

Manning, A. and Dawkins, M.S. (1992) *An Introduction to Animal Behaviour* (4th edn). Cambridge: Cambridge University Press.

Menzel, E.W. (1978) Cognitive mapping in chimpanzees. In S.H. Hulse, H. Fowler and W.K. Honig (eds), *Cognitive Processes in Animal Behaviour*. Hillsdale, N.J.: Erlbaum.

Mittelstaedt, H. and Mittelstaedt, M.L. (1982) Homing by path integration. In F. Papi and H.G. Wallraff (eds), *Avian Navigation*. New York: Springer-Verlag.

Morgan, C.L. (1894) *An Introduction to Comparative Psychology*. London: Scott.

Morris, R.G.M., Garrud, P., Rawlins, J.N.P. and O'Keefe, J. (1982) Place navigation impaired in rats with hippocampal lesions. *Nature*, 297, 681–683.

Moynihan, M.H. and Rodaniche, A.F. (1977) Communication, crypsis, and mimicry among cephalopods. In T.A. Sebeok (ed.), *How Animals Communicate*. Bloomington, Ind.: Indiana University Press.

Muller-Schwarze, D. (1971) Pheromones in black-tailed deer (*Odocoileus hemionus columbianus*). *Animal Behaviour*, 19, 141–152.

O'Keefe, J. and Speakman, A. (1987) Single unit activity in the rat hippocampus during a spatial memory task. *Experimental Brain Research*, 68, 1–27.

Olton, D.S., Collison, C. and Werz, M.A. (1977) Spatial memory and radial arm maze performance in rats. *Learning and Motivation*, 8, 289–314.

Olton, D.S. and Samuelson, R.J. (1976) Remembrance of places passed: Spatial memory in rats. *Journal of Experimental Psychology: Animal Behaviour Processes*, 2, 97–116.

Overton, D.A. (1964) State-dependent or 'dissociated' learning produced with pentobarbitol. *Journal of Comparative and Physiological Psychology*, 57, 3–12.

Papi, F. and Luschi, P. (1996) Pinpointing 'Isla Meta': The case for sea turtles and albatrosses. *Journal of Experimental Biology*, 199, 65–71.

Papi, F., Luschi, P. and Limonta, P. (1992) Orientation-disturbing magnetic treatment affects the pigeon opiod system. *The Journal of Experimental Biology*, 166, 169–179.

Patterson, F.G. (1978) Linguistic capabilities of a lowland gorilla. In F.C.C. Peng (ed.), *Sign Language and Language Acquisition in Man and Ape: New dimensions in comparative pedolinguistics*. Boulder, Colo.: Westview Press.

Patterson, F.G. (1980) Innovative uses of language by a gorilla: A case study. In K. Nelson (ed.), *Children's Language* Vol. 2. New York: Gardner Press.

Patterson, F.G. and Linden, E. (1981) *The Education of Koko*. New York: Holt, Rinehart and Winston.

Pearce, J.M. (1997) *Animal Learning and Cognition* (2nd edn). Hove: Psychology Press.

Perdeck, A.C. (1958) Two types of orientation in migrating starlings, *Sturnus vulgaris* L., and chaffinches, *Fringilla coelebs* L., as revealed by displacement experiments. *Ardea*, 46, 1–37.

Peters, R.P. and Mech, L.D. (1975) Scent marking in wolves. *American Scientist*, 63, 628–637.

Pfungst, O. (1965) *Clever Hans: The horse of Mr Van Osten*. New York: Holt.

Poucet, B. (1993) Spatial cognitive maps in animals, new hypotheses on their structure and neural mechanisms. *Psychological Review*, 100, 163–182.

Premack, D. (1971) Language in a chimpanzee? *Science*, 172, 808–822.

Premack, D. (1976) *Intelligence in Ape and Man*. Hillsdale, N.J.: Lawrence Erlbaum Associates.

Renner, M. (1960) Contribution of the honey bee to the study of time sense and astronomical orientation. *Cold Spring Harbor Symposium on Quantitative Biology*, 25, 361–367.

Ridley, M. (1995) *Animal Behaviour: A concise introduction* (2nd edn). Oxford: Blackwell.

Roberts, W.A. (1992) Foraging by rats on a radial maze. Learning, memory and decision rules. In I. Gormezano and E.A. Wasserman (eds), *Learning and Memory: The behavioral and biological substrates*. Hillsdale, N.J.: Erlbaum.

Roberts, W.A. (1998) *Principles of Animal Cognition*. Boston, Mass.: McGraw-Hill.

Rumbaugh, D.M. (1977) *Language Learning by a Chimpanzee: The LANA project*. New York: Academic Press.

Rumbaugh, D.M. and Savage-Rumbaugh, E.S. (1994) Language in comparative perspective. In N.J. Mackintosh (ed.), *Animal Learning and Cognition*. London: Academic Press.

Saint Paul, U.V. (1982) Do geese use path integration for walking home? In F. Papi and H.G. Wallraff (eds), *Avian Navigation*. New York: Springer-Verlag.

Savage-Rumbaugh, E.S. (1986) *Ape Language: From conditioned response to symbol*. New York: Columbia University Press.

Savage-Rumbaugh, E.S. and Brakke, K.E. (1996) Animal language: Methodological and interpretive issues. In M. Bekoff and D. Jamieson (eds), *Readings in Animal Cognition*. Cambridge, Mass.: The MIT Press.

Savage-Rumbaugh, E.S. and Lewin, R. (1994). *Kanzi: The ape at the brink of the human mind*. New York: John Wiley and Sons.

Savage-Rumbaugh, E.S., Murphy, J., Sevcik, R.A., Brakke, K.E., Williams, S.L. and Rumbaugh, D.M. (1993) Language comprehension in ape and child. *Monographs of the Society for Research in Child Development*, 58, 1–256.

Schneider, D. (1974) The sex-attractant receptor of moths. *Scientific American*, 231, 28–35.

Seyfarth, R.M. and Cheney, D.L. (1986) Vocal development in vervet monkeys. *Animal Behaviour*, 34, 1640–1658.

Seyfarth, R.M., Cheney, D.L. and Marler, P. (1980) Monkey responses to three different alarm calls: Evidence of predator classification and semantic communication. *Science*, 210, 801–803.

Shanker, S.G., Savage-Rumbaugh, E.S. and Taylor, T.J. (1999) Kanzi: A new beginning. *Animal Learning and Behaviour*, 27, 24–26.

Sherry, D.F. (1992) Landmarks, the hippocampus, and spatial food search in food-storing birds. In W.K. Honig and J.G. Fetterman (eds), *Cognitive Aspects of Stimulus Control*. Hillsdale, N.J.: Erlbaum.

Sherry, D.F. and Duff, S.J. (1996) Behavioural and neural bases of orientation in food-storing birds. *Journal of Experimental Biology*, 199, 165–172.

Sherry, D.F., Forbes, M.R.L., Khurgel, M. and Ivy, G.O. (1993) Females have a larger hippocampus than males in the brood-parasitic brown-headed cowbird. *Proceedings of the National Academy of Sciences*, 90, 7839–7843.

Sherry, D.F. and Vaccarino, A.L. (1989) Hippocampus and memory for food caches in black-capped chickadees. *Behavioural Neuroscience*, 103, 308–318.

Shettleworth, S.J. (1998) *Cognition, Evolution and Behavior*. New York: Oxford University Press.

Shorey, H.H. (1976) *Animal Communication by Pheromones*. New York: Academic Press.

Smith, W.J. (1996) Communication and expectations: A social process and the cognitive operations it depends upon and influences. In M. Bekoff and D. Jamieson (eds), *Readings in Animal Cognition*. Cambridge, Mass.: The MIT Press.

Terrace, H.S. (1979) *Nim*. New York: Knopf.

Terrace, H.S., Petitto, L.A., Sanders, R.J. and Bever, T.G. (1979) Can an ape create a sentence? *Science*, 206, 891–902.

Thorpe, W.H, (1961) *Bird Song*. Cambridge: Cambridge University Press.

Tinbergen, N. (1951) *The Study of Instinct*. Oxford: Oxford University Press.

Tolman, E.C. (1948) Cognitive maps in rats and men. *Psychological Review*, 55, 189–208.

Tolman, E.C. and Honzik, C.H. (1930) 'Insight' in rats. *University of California Publications in Psychology*, 4, 215–232.

Tomasello, M. and Call, J. (1997) *Primate Cognition*. New York: Oxford University Press.

Tomback, D.F. (1980) How nutcrackers find their seed stores. *Condor*, 82, 10–19.

Tyack, P. (1983) Differential response of humpback whales *Megaptera novaengliae* to playback of song or social sounds. *Behavioural Ecology*, 13 (1), 49–55.

Van Beusekom, G. (1948) Some experiments on the optical orientation in *Philanthus triangulum. Behaviour*, 1, 195–225.

Vauclair, J. (1990) Primate cognition: From representation to language. In S.T. Parker and K.R. Gibson (eds), *'Language' and Intelligence in Monkeys and Apes: Comparative Developmental Perspectives*. New York: Cambridge University Press.

Vauclair, J. (1996) *Animal Cognition*. Cambridge, Mass.: Harvard University Press.

Von Frisch, K. (1950) *Bees, their Vision, Chemical Senses, and Language*. Oxford: Oxford University Press.

Von Frisch, K. (1962) Dialects in the language of bees. *Scientific American*, 207, 78–87.

Von Frisch, K. (1974) Decoding the language of the bee. *Science*, 185, 663–668.

Walcott, C. and Schmidt-Moenig, K. (1973) The effect of anaesthesia during displacement on the homing performance of pigeons. *Auk*, 90, 281–286.

Wallman, J. (1992) *Aping Language*. Cambridge: Cambridge University Press.

Wallraff, H.D. (1984) Migration and navigation in birds: a present-state survey with some digressions to related fish behaviour. In J.D. McCleave (ed.), *Mechanisms of Migration in Fishes*. New York: Plenum.

Wallraff, H.G. (1990) Navigation by homing pigeon. *Ethology, Ecology and Evolution*, 2, 81–115.

Waterman, T.H. (1989) *Animal Navigation*. New York: Scientific American Press.

Wehner, R. (1992) Arthropods. In F. Papi (ed.), *Animal Homing*. London: Chapman and Hall.

Wenner, A.M. and Wells, P.H. (1990) *Anatomy of a Controversy*. New York: Columbia University Press.

Wilson, E.O. (1965) Chemical communication in the social insects. *Science*, 149, 1064–1071.

Wiltschko, W. and Wiltschko, R. (1996) Magnetic orientation in birds. *Journal of Experimental Biology*, 199, 29–38.

Yodlowski, M.L., Kreithen, M.L. and Keeton, W.T. (1977) Detection of atmospheric infrasound by homing pigeons. *Nature* (London) 265, 725–726.

Index